SpringerBriefs in Electrical and Computer Engineering

Signal Processing

More information about this series at http://www.springer.com/series/11560

Chen Chen · Yuzhuo Ren · C.-C. Jay Kuo

Big Visual Data Analysis

Scene Classification and Geometric Labeling

Springer

Chen Chen
Department of Electrical Engineering
University of Southern California
Los Angeles, CA
USA

C.-C. Jay Kuo
University of Southern California
Los Angeles, CA
USA

Yuzhuo Ren
Department of Electrical Engineering
University of Southern California
Los Angeles, CA
USA

ISSN 2191-8112 ISSN 2191-8120 (electronic)
SpringerBriefs in Electrical and Computer Engineering
ISSN 2196-4076 ISSN 2196-4084 (electronic)
SpringerBriefs in Signal Processing
ISBN 978-981-10-0629-6 ISBN 978-981-10-0631-9 (eBook)
DOI 10.1007/978-981-10-0631-9

Library of Congress Control Number: 2016932343

Printed on acid-free paper

This Springer imprint is published by Springer Nature
The registered company is Springer Science+Business Media Singapore Pte Ltd.

Dedicated to my wife and my parents, for their love and endless support

—Chen Chen

Dedicated to my parents for their endless love and encouragement

—Yuzhuo Ren

Dedicated to my wife for her long-term understanding and support

—C.-C. Jay Kuo

Preface

Scene understanding is a key issue in computer vision, which recognizes scene image semantic contents and their corresponding contexts. As one of the most challenging scene understanding problems, scene classification considers the semantic concepts in a scene image and classifies scene images into their associated scene categories. Meanwhile, geometric labeling focuses on the scene layouts, where image pixels are labeled and grouped into different geometrically functional classes. It is also an essential step before any further scene understanding tasks such as recognition, annotation, and retrieval, and its results affect the performance of these applications significantly. With rapidly increasing visual data in volume, variety and velocity, traditional approaches toward solutions of the two problems are not adequate in addressing the new challenges. In this brief, we attempt to provide more accurate and scalable solutions for several scene classification and geometric labeling problems: indoor/outdoor classification, outdoor scene classification, and scene layout estimation. This brief will first give an overview on the state-of-the-art algorithms for each problem. Then, it will introduce several techniques for corresponding solutions. They are Expert Decisions Fusion (EDF), Coarse Semantic Segmentation (CSS) and Global-attributes Assisted Labeling (GAL). Extensive experimental results with comparative analysis will be provided for each approaches. Finally, we will conclude and highlight the contributions of these approaches to big visual data analysis.

Contents

Chapter 1
Introduction

Keywords Big visual data · Indoor/Outdoor classification · Outdoor scene classification · Geometric labeling · Decision fusion · Structured machine learning system · Contour-guided color palette · Semantic segmentation · Context-aware labeling · Global attributes

Visual data analysis has been one of the most popular research fields in the past two decades. In literature, topics such as detection, tracking, matching, recognition, tagging, classification and retrieval are the most frequently visited ones, and all of them can be categorized into the task of visual understanding. The general research focus has been on developing solutions that can accurately and robustly understand semantic meanings of visual data, which includes both 2-D images and videos. However, many existing solutions are not suitable for real-world visual data analysis applications because of their limited generalities. Recently, with the emergence of more research resources (e.g., datasets), the field of visual data analysis has evolved with a huge progress. In this book, we will focus on the most up-to-date research works in big visual data analysis. More specifically, we will discuss classification and labeling algorithms, which provides the state-of-the-art visual understanding performance in indoor/outdoor scene classification, outdoor scene categorization and outdoor scene geometric labeling. To demonstrate the progresses made by these algorithms, we will firstly show the inadequacy of traditional approaches when dealing with large-scale visual data. We will see the big visual data analysis challenges, such as scalability and generality. Then corresponding to different problems created by big visual data, three state-of-the-art algorithms will be introduced. They include Expert Decision Fusion (EDF), Coarse Semantic Segmentation (CSS), and Global-attributes Assisted Labeling (GAL). EDF offers a systematic strategy to improve the performance of any "understanding" task of big visual data by taking advantages of multiple existing experts in a well defined structured system. CSS provides semantic and accurate local-scale machine learning units that can be used in semantic-based image representations. Finally, GAL introduces several generalized global attributes which are significant to machine learning process in the sense of global recognition.

C. Chen et al., *Big Visual Data Analysis*,
SpringerBriefs in Signal Processing, DOI 10.1007/978-981-10-0631-9_1

As a very significant problem in visual data analysis, scene understanding targets at automatically extracting semantic and valuable descriptions of given images and videos. Scene classification and scene geometric labeling are both popular topics in scene understanding. In scene classification, Indoor/outdoor classification aims at categorizing scene images into indoor and outdoor groups. More generally, outdoor scene categorization aims at classifying outdoor scene images into several semantic classes such as coast, street and beach. Different from classification tasks, geometric labeling focuses on local recognition and representation of a semantic regions in the given visual data. It provides geometrically functional labels to regions such as sky (farthest vertical plane), ground and water (horizontal planes) and building facades (orientated vertical surfaces). As a result, geometric labeling is especially significant to scene layout estimation related problems.

Traditionally, the design of good features (from low-level to high-level), which is called feature extraction, plays the most critical role in addressing these problems. Then, it is followed by the design of different classifiers with tools such as SVM [2], Adaboost [7] and logistic regression [1]. In the field of scene classification, quite a few low-level features have been developed to achieve impressive results in object detection [14, 23], pedestrian detection [5, 26] and scene recognition [21, 24]. For example, the HOG descriptor [4] is the normalized histogram of gradient orientations in a local region, and it is widely used in object and pedestrian detection. GIST [20] measures the scene global characteristics such as openness, toughness and crowdedness, by local edge orientation responses, and it is widely used in scene classification. Oliva et al. [20] reported a nearly 85 % classification accuracy in an eight-scene classification tasks using GIST. The SIFT descriptor [17] is a popular local feature that is invariant to scale and rotation. SIFT finds its applications in object detection [6], object tracking [27], image retrieval [3, 15] and scene categorization [22]. Most scene understanding solutions are based on these low-level features. However, the performance of these state-of-the-art solutions is far below that of the human vision system (HVS).

To better interpret an scene, methods based on low-level features attempt to capture local characteristics of image pixels. Recently, features that model the global image content are also developed. The "Bag of Words (BoW)" model [11, 12] is a document representation of an image, which captures the statistics of local features in an image. However, due to the high diversity of image contents, the performance of the BoW model is still far behind the HVS [13, 18, 19]. The restricted representation power of features and the simplicity of the feature models mainly come from the limited capability on feature extraction units (pixels, blocks and super-pixels), which prevents them from achieving desirable performance in a large-scale database. Xiao et al. [25] tested twelve different features in scene classification tasks on SUN database. The averaged correct classification rates of individual features range from less than 5 to 40 % for a single scene. Although it is possible to combine all features into a long feature vector, the accuracy is not significantly improved. Besides, to control the training cost, the number of images should be significantly reduced in the training process. It is well known in machine learning that a long feature vector demands a higher computational cost and may lead to over-fitting.

Other than the significant problem in feature extraction, lacking a comprehensive global description of scene images is also a key issue in most understanding tasks. Local descriptions of segments color, texture and line orientations have been well studied in [10]. However, researchers quickly recognize the limitations and started to explore more global representations using 3-D block models [8, 9] and predefined geometric rules [16]. Unfortunately, these models and rules require strong assumptions on data types. When we have diverse image types, performance of these approaches become miserable.

The indoor/outdoor classification and the outdoor scene classification and geometric labeling problems face all these challenges as described above. To equip reader with strong knowledge about the state-of-the-art big visual data analysis problems and solutions, this book provides a comprehensive survey on these three important and popular topics. To show that different low-level feature models, which are called experts in the book, have different strengths for different kinds of data, we propose an EDF technique to integrate the opinions of different experts. We will demonstrate that an EDF system can perform more accurately and robustly against a big visual dataset. To present the significance of reasonable learning units in visual understanding tasks, we propose a CSS technique to achieve semantic and robust local learning units for accurate representations of scene images in classification tasks. CSS outperforms all existing algorithms for a very challenging dataset in outdoor categorization. Finally, to take advantage of global understanding process in human visual system, we introduce GAL to integrate the most significant global attributes into machine learning systems. In solving the geometric labeling problem, GAL presents a much superior performance and generality comparing to existing works.

The rest of this book is organized as following. We introduce some state-of-the art visual datasets to demonstrate the challenges in big visual data understanding problems in Chap. 2. We give an overview on indoor/outdoor classification, and we introduce EDF for indoor/outdoor classification in Chap. 3. In Chap. 4, we will review the outdoor scene classification problem and traditional solutions. We will then introduce the proposed CSS algorithm and visualize the robust semantic learning units for superior classification performance. In Chap. 5, as the algorithm that handling global representation of scene images, we will introduce the GAL, which solves the geometric labeling problem and provides the state-of-the-art performance. Finally, concluding remarks on big visual data analysis, the introduced algorithms and future research topics are given in Chap. 6.

References

1. Boyd, C.R., Tolson, M.A., Copes, W.S.: Evaluating trauma care: the triss method. J. Trauma-Inj., Infect., Crit. Care **27**(4), 370–378 (1987)
2. Chang, C.C., Lin, C.J.: Libsvm: a library for support vector machines. ACM Trans. Intell. Syst. Technol. (TIST) **2**(3), 27 (2011)

3. Chatzichristofis, S.A., Boutalis, Y.S.: Cedd: color and edge directivity descriptor: a compact descriptor for image indexing and retrieval. In: Computer Vision Systems, pp. 312–322. Springer (2008)
4. Dalal, N., Triggs, B.: Histograms of oriented gradients for human detection. In: IEEE Computer Society Conference on Computer Vision and Pattern Recognition. CVPR 2005, vol. 1, pp. 886–893 (2005)
5. Dollar, P., Wojek, C., Schiele, B., Perona, P.: Pedestrian detection: an evaluation of the state of the art. IEEE Trans. Pattern Anal. Mach. Intell. **34**(4), 743–761 (2012)
6. Felzenszwalb, P.F., Girshick, R.B., McAllester, D., Ramanan, D.: Object detection with discriminatively trained part-based models. IEEE Trans. Pattern Anal. Mach. Intell. **32**(9), 1627–1645 (2010)
7. Freund, Y., Schapire, R.E., et al.: Experiments with a new boosting algorithm. In: ICML, vol. 96, pp. 148–156 (1996)
8. Gupta, A., Efros, A.A., Hebert, M.: Blocks world revisited: image understanding using qualitative geometry and mechanics. In: Computer Vision ECCV 2010, pp. 482–496. Springer (2010)
9. Gupta, A., Hebert, M., Kanade, T., Blei, D.M.: Estimating spatial layout of rooms using volumetric reasoning about objects and surfaces. In: Advances in Neural Information Processing Systems, pp. 1288–1296 (2010)
10. Hoiem, D., Efros, A., Hebert, M., et al.: Geometric context from a single image. In: Tenth IEEE International Conference on Computer Vision. ICCV 2005, vol. 1, pp. 654–661. IEEE (2005)
11. Jégou, H., Douze, M., Schmid, C.: Improving bag-of-features for large scale image search. Int. J. Comput. Vis. **87**(3), 316–336 (2010)
12. Lazebnik, S., Schmid, C., Ponce, J.: Beyond bags of features: spatial pyramid matching for recognizing natural scene categories. In: Computer Vision and Pattern Recognition, 2006 IEEE Computer Society Conference, vol. 2, pp. 2169–2178 (2006)
13. Li, L.J., Fei-Fei, L.: What, where and who? classifying events by scene and object recognition. In: IEEE 11th International Conference on Computer Vision. ICCV 2007, pp. 1–8. IEEE (2007)
14. Lienhart, R., Maydt, J.: An extended set of haar-like features for rapid object detection. In: 2002 International Conference on Image Processing. Proceedings., vol. 1, pp. I–900. IEEE (2002)
15. Lim, J.H., Jin, J.S.: A structured learning framework for content-based image indexing and visual query. Multimed. Syst. **10**(4), 317–331 (2005)
16. Liu, X., Zhao, Y., Zhu, S.C.: Single-view 3D scene parsing by attributed grammar. In: 2014 IEEE Conference on Computer Vision and Pattern Recognition (CVPR), pp. 684–691. IEEE (2014)
17. Lowe, D.G.: Distinctive image features from scale-invariant keypoints. Int. J. Comput. Vis. **60**(2), 91–110 (2004)
18. Niebles, J.C., Fei-Fei, L.: A hierarchical model of shape and appearance for human action classification. In: IEEE Conference on Computer Vision and Pattern Recognition. CVPR'07, pp. 1–8. IEEE (2007)
19. Niebles, J.C., Wang, H., Fei-Fei, L.: Unsupervised learning of human action categories using spatial-temporal words. Int. J. Comput. Vis. **79**(3), 299–318 (2008)
20. Oliva, A., Torralba, A.: Modeling the shape of the scene: a holistic representation of the spatial envelope. Int. J. Comput. Vis. **42**(3), 145–175 (2001)
21. Quattoni, A., Torralba, A.: Recognizing indoor scenes. In: Computer vision and pattern recognition (CVPR), 2009 IEEE conference (2009)
22. van Gemert, J.C., Geusebroek, J.M., Veenman, C.J., Smeulders, A.W.: Kernel codebooks for scene categorization. In: Computer Vision-ECCV 2008, pp. 696–709. Springer (2008)
23. Viola, P., Jones, M.: Rapid object detection using a boosted cascade of simple features. In: Proceedings of the 2001 IEEE Computer Society Conference on Computer Vision and Pattern Recognition. CVPR 2001, vol. 1, pp. I–511. IEEE (2001)
24. Wu, J., Rehg, J.M.: Centrist: a visual descriptor for scene categorization. IEEE Trans. Pattern Anal. Mach. Intell. **33**(8), 1489–1501 (2011)

25. Xiao, J., Hays, J., Ehinger, K.A., Oliva, A., Torralba, A.: Sun database: large-scale scene recognition from abbey to zoo. In: Computer Vision and Pattern Recognition (CVPR), 2010 IEEE Conference, pp. 3485–3492. IEEE (2010)
26. Yan, J., Zhang, X., Lei, Z., Liao, S., Li, S.Z.: Robust multi-resolution pedestrian detection in traffic scenes. In: 2013 IEEE Conference on Computer Vision and Pattern Recognition (CVPR), pp. 3033–3040. IEEE (2013)
27. Zhou, H., Yuan, Y., Shi, C.: Object tracking using sift features and mean shift. Comput. Vis. Image Underst. **113**(3), 345–352 (2009)

Chapter 2
Scene Understanding Datasets

Keywords Dataset · Large-scale · PASCAL dataset · ImageNet · LabelMe dataset · Fifteen scene category dataset · CMU 300 dataset · Tiny image dataset · SUN dataset · PLACE205 dataset

2.1 Small-Scale Scene Understanding Datasets

At early ages of scene understanding, several benchmarks were proposed for research purposes. The 8-scene dataset, the 15-scene dataset, the UIUC sport dataset were dominate ones. These traditional datasets have several common properties. First, they consist of typical and iconic scene images that background objects and surfaces are the determinative components to decide scene categories. Besides, these datasets have limited number of scene categories and images within the same class have little variations in visual patterns. Finally, they focus more on usually-seen categories, such as "coast" and "living room", etc., and do not have seldom-seen scenes or detailed categories such as "temple", "auditorium", etc.

2.1.1 8-scene Dataset

8-scene dataset was firstly used in the work [1] where the famous image descriptor "GIST" was proposed. It consists of 8 outdoor scene categories (4 of them are from natural landscapes and 4 of them are from man-made scenes). Totally, there are 2688 images in the dataset. It has been treated as the most famous benchmark in the scene understanding field and a mandatory challenge for all scene understanding researches since then. Later, researchers from the same group labeled the dataset for

C. Chen et al., *Big Visual Data Analysis*,
SpringerBriefs in Signal Processing, DOI 10.1007/978-981-10-0631-9_2

Fig. 2.1 Examples in 8-scene dataset

the purposes of semantic segmentation in scene parsing researches. The segmentally labeled data was usually called SIFTFlow after the work [2]. Examples images from the 8 scenes and corresponding segmental labels can be seen in Fig. 2.1.

2.1.2 15-scene Dataset

After the blooming of 8-scene [1], researchers intended to increase the categories and diversities to create more challenging datasets. Inheriting the popularity of the 8-scene dataset, 15-scene dataset includes 2 extra outdoor categories and 5 indoor categories to meet the requirement on general scene understanding tasks. As shown in Fig. 2.2, proposed by [3, 4], the 15-scene dataset has three more challenging factors than the 8-scene datasets: (1) the 15-scene dataset consists of only gray-scale images, (2) the 15-scene dataset has twice number of scene categories and images and (3) the 15-scene dataset considers indoor scenes with outdoor scenes together. With these challenges and differences, bonded with the 8-scene dataset, the 15-scene dataset also became a mandatory benchmark in nowadays researches.

Fig. 2.2 Examples in 15-scene dataset: stared categories are originally from the 8-scene dataset

2.1.3 UIUC Sports

UIUC sports [5] contains 8 sports event categories: rowing (250 images), badminton (200 images), polo (182 images), bocce (137 images), snowboarding (190 images), croquet (236 images), sailing (190 images), and rock climbing (194 images). Images are divided into easy and medium according to the human subject judgment. Information of the distance of the foreground objects is also provided for each image. Unlike the 8-scene dataset and the 15-scene dataset, UIUC sports is an event-centric scene dataset. As shown in Fig. 2.3, most of its scene images have distinctive foreground and background contexts. Therefore it arouses a promising research direction in context based recognition using scene/object topic models.

2.1.4 CMU 300

The dataset in [6] is the largest benchmarking dataset in the geometric layout research community. It consists of 300 images of outdoor scenes with 23 different scene categories including alley, building, cliff, college etc. Example images are shown in Fig. 2.4. The dataset provides ground truth geometric labels for each image, namely support, sky, planar left, planar center, planar right, non-planar solid and porous. The first 50 images are used for training the surface segmentation algorithm as done in previous work [6, 7]. The remaining 250 images are used for evaluation. Besides geometric labels, it also provides occlusion boundaries for 100 images in this dataset. The occlusion boundaries indicate the depth orders of occluding and occluded objects.

Fig. 2.3 Examples in UIUC sports dataset

Fig. 2.4 Examples in CMU 300 dataset

2.2 Large-Scale Scene Understanding Datasets

With progresses made in computer vision research, small traditional image datasets are no longer sufficient for the performance evaluation purpose of robust scene understanding system, and several large scale datasets are built to meet the urgent need. We will present several of them commonly used for scene understanding.

2.2.1 *80 Million Tiny Image Dataset*

The 80 million tiny image dataset contains 7,527,697 images. It was built by [8]. Each image is of low resolution (with image size dimension 32 × 32) and labeled with one of the 53,464 English nouns from the WordNet [9]. It is mainly used in fast image search which demands very little memory. All images were obtained from the Google search and other engines using English nouns from the WordNet. These images are with high diversity, containing object images and scene images.

The construction of this dataset was motivated by an interesting experimental observation. That is, human can classify a scene with 32 × 32 pixels and achieve a recognition rate higher than 80 % [8]. The purpose of creating this dataset is to facilitate the development of fast image search and scene matching techniques with very little memory. It combines Vogel and Schiele [10] 702 natural scenes, Olivia and Torralba's [1] 2688 images, Caltech 101 categories, Caltech 256 categories. The Caltech 101 categories and Caltech 256 categories are images containing objects widely used in object recognition. The 80 million tiny image dataset contains image categories of sufficient diversity. Although the image number is large, all images are categorized.

In the project website [8], it provides the confidence map, labels provided by users, nouns from the WordNet and the visual dictionary view of the whole dataset. The confidence map reflects the algorithmic accuracy in the classification of different categories. The labels show the amount of data provided by users. It also offers the mosaic image after images are divided into semantic groups. An example can be seen in Fig. 2.5. The visual dictionary view provides the visualization of tiles by averaging the color of images that have the same English noun. The averaged version of a class of images can reflect the global information of this class. The website allows users to add annotations by selecting a word and indicating whether correct and wrong

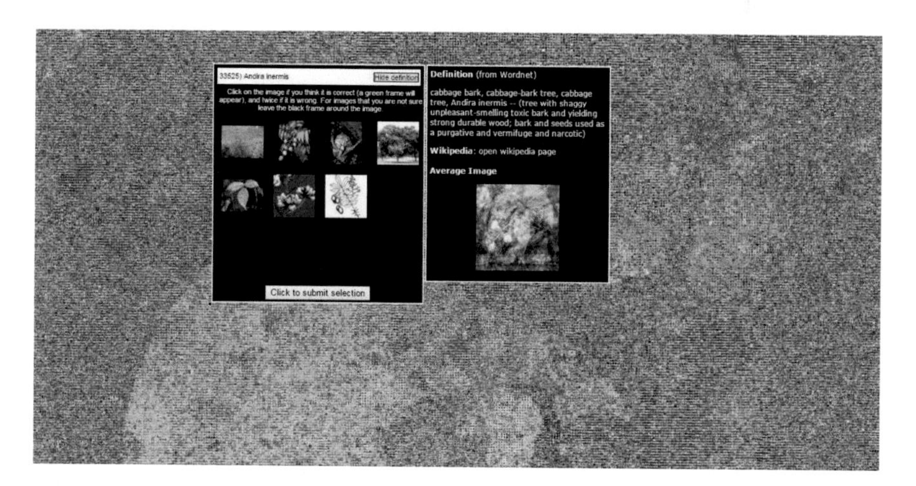

Fig. 2.5 The poster of the visual dictionary built in the tiny image dataset

images are returned. To fix inaccurate annotations, it allow users to correct labels online. However, the low resolution of these tiny images usually do not allow image classification algorithms to work properly.

2.2.2 PASCAL Dataset

The PASCAL Visual Object Classes (VOC) Challenge dataset [11] provides standardized image datasets for object class recognition and a common set of tools to access the data sets and annotations. There are 20 object classes in the PASCAL dataset with thousands of images in each class. Examples of object images can be seen in Fig. 2.6. The images were obtained from the collection of Flicker photos. The PASCAL VOC Challenge was an annual event from 2005 to 2012, in which researchers submitted their results on object classification, and got their results evaluated and compared online. The competition includes:

- Classification: determining presence/absence of an example of that class in the test image for each of the 20 classes.
- Detection: determining the bounding box and the label of each object from the 20 target classes in the test image.
- Segmentation: generating the pixel-wise segmentation of an object in the image.
- Person layout: determining the bounding box and the label of each part of a person.
- Action classification: determining the action of a person in a still image.

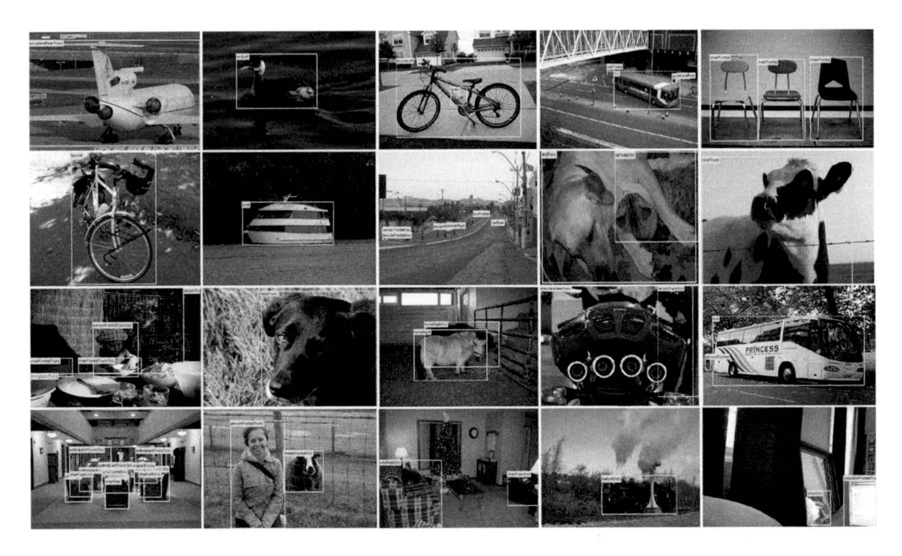

Fig. 2.6 The 20 object classes in PASCAL dataset

Although the PASCAL dataset is a large and challenging dataset for object classification and recognition, it is not an appropriate dataset for our interest in dealing with scene images.

2.2.3 *ImageNet Dataset*

The ImageNet [12] is an image dataset organized according to the WordNet hierarchy. Each meaningful concept in the WordNet, possibly described by multiple words or word phrases, is called a "synonym set" or "synset". There are more than 100,000 synsets in WordNet, and a great majority of them are nouns (80,000+). The ImageNet aims to provide on average 1000 images to illustrate each synset. Images of each concept are quality-controlled and human-annotated. Compared to the other image classification dataset, the ImageNet is the largest and most challenging dataset for object classification and recognition. On the other hand, it focuses on general image classification challenges, which include scene classification as only a small branch of the problem. Figure 2.7 shows several examples in the dataset. We can see most of them are foreground objects oriented.

2.2.4 *LabelMe Dataset*

LabelMe [13] is a project conducted by the MIT CSAIL with an objective to provide a dataset of digital images with object and surfaces annotations. The dataset is dynamic, free to use, and open to public contribution. Specifically, it provides a website for people to annotate images online. An example can be seen in Fig. 2.8. LabelMe asks users to use polygons to segment and annotate object and surfaces in an image. As of October 31, 2010, LabelMe has 187,240 images, 62,197 annotated images, and 658,992 labeled objects. The project was motivated by the following observation. Most available data in computer vision research are tailored to the problem of a specific research group and it is often that new researchers need to collect additional data to solve their own problems. LabelMe was created to solve this shortcoming. LabelMe are different from other existing datasets in the following aspects. First, LabelMe contains images of objects with multiple angles, sizes and orientations. Second, LabelMe designs images for object recognition in arbitrary scenes and it avoids the scene to cropped, normalized or resized. Third, each image in LabelMe may contain more than one object, and users are allowed to label these objects. Finally, its numbers of images and object classes can be easily increased.

Fig. 2.7 Examples in the ImageNet dataset

2.2.5 *Scene Understanding (SUN) Dataset*

The Scene understanding (SUN) dataset, introduced by Xiao et al., finds applications in many research fields, such as scene recognition, computer vision, human perception, cognition and neuroscience, machine learning, data mining, computer graphics and robotics research. The SUN dataset contains 899 categories and 130,519 images [14] in total. It has been widely used in various scene related computer vision researches, such as scene classification [14], scene recognition [15], and indoor/outdoor image classification [16]. It was motivated by the demand to build a rich and diverse dataset that includes our daily experienced scenes in the real world as much as possible. Being different from the object detection datasets such as the PASCAL [11] and the Caltech 256 category datasets [17], images in the SUN dataset are all about scenes where human can navigate or interact with. The SUN dataset

Fig. 2.8 An example of annotations in LabelMe

is currently the largest scene dataset in terms of the image number and the scene number.

The scene category of the SUN dataset is huge. We can easily think of some scene categories such as the coast, the field, the meeting room etc. Is Grand Canyon a scene? Should it be a category? How can we include as many scene terms as possible? The category terms are selected from the 70,000 terms of the WordNet [9] used in the tiny images dataset [8]. These terms describe scenes, places and environments. There are several criteria in selecting scene category terms. First, places terms that are too broad to evoke a specific visual identity (such as territory, workplace and outdoors) and places names (such as Grand Canyon or New York) are not included. Second, specific types of objects which are scene related are included, such as buildings (skyscraper, house and hangar) which makes the scene categories more diverse. Third, it contains specific domains such as the pine forest, rainforest and orchard which all belong to the wooded area. Besides the terms in the WordNet, a few categories missed by the WordNet are added. After the first round of term selection, there are about 2500 initial terms. After merging terms by synonyms and separating scenes of different visual identities (indoor and outdoor views of churches), there are 899 categories. This is far more than the previously created 8-scene dataset [1] and the 15-scene dataset [4]. We can see that the SUN dataset really contains comprehensive scene categories in Fig. 2.9.

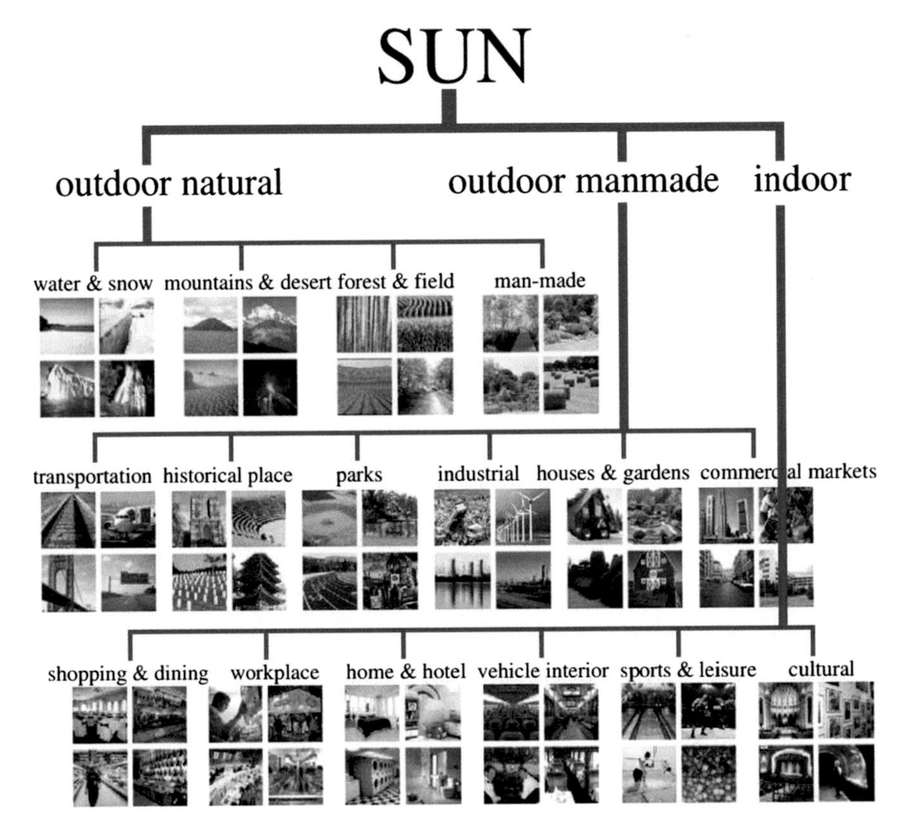

Fig. 2.9 Visualization of scene hierarchies in SUN

There is a subset of the SUN 899 categories dataset containing 397 categories. It is more popular in the research field since each category contains at least 100 images. Other categories not included in the 397 category contain fewer images. The total number in the SUN database is large. Images in the SUN dataset come from the Internet search. In each category of the 8-scene and the 15-scene dataset contains hundreds of images, which is far less than that in the SUN dataset. In contrast with the eight scene and the fifteen scene datasets, images in each category of the SUN dataset are more diversified. As a result, this dataset imposes more challenges on the scene classification and recognition tasks.

As compared with the 80 million tiny image dataset, images in the SUN dataset are of much higher resolution. Images in the SUN dataset have a resolution of at least 200×200 pixels. Degenerate or unusual images (black and white, distorted colors, very blurry or noisy, incorrectly rotated, aerial views, noticeable borders) were removed in the image collection process.

[14] conducted experiments to compare human scene classification accuracy and machine classification accuracy. The purpose was to show that the SUN dataset was

constructed consistently with minimal overlap between categories and that the scene classification task would be a difficult one.

To facilitate participants to understand the 397 scene categories, category terms are grouped in a three-level tree. The category on the leaf-level is the most specific one. The participants can navigate through the three-level hierarchy tree to reach a specific scene type. Each leaf-level SUN category interface shows an example of the image to help participants better know what the image looks like in each category.

Human scene classification accuracy was measured on 20 distinct test scenes in each category by Amazon's Mechanical Turk (AMT). There are $397 \times 20 = 7940$ experiments or HITs (Human Intelligence Tasks in AMT parlance). Human take 61 s per experiment and can achieve 58.6 % accuracy at the leaf level on the average. Considering the large number of categories, human classification accuracy is already very high. There are "good works" that can achieve accuracy as high as 95 % on the relatively easy first level of the hierarchy and the leaf-level accuracy rises to 68.5 %. One author involved in constructing the SUN dataset achieved 97.5 % at the first-level and 70.6 % at the leaf-level. The "good works" are trustworthy. By analyzing the confusion scene categories, these confused scenes are semantically similar and they are restricted to only a few scenes. The experiment shows that human classification accuracy is far below 100 % at the leaf-level. Image classification is actually a difficult task for human, and it is even more difficult for the computer.

Twelve individual methods are compared in scene classification, including GIST [1], HOG2x2 [18], Dense SIFT [4], LBP [19], Sparse SIFT histograms [20, 21], SSIM [20], Tiny Images [8], Line Features [22, 23], Texton Histograms [24], Color Histograms, Geometric Probability Map [25] and Geometric specific histograms [26]. The extracted features are all relevant to scene classification. The classifier is trained using the one-vs-remaining SVM. To compare the performance on different datasets, experiments are also conducted on the 15-scene dataset [1, 3, 4]. The results are shown in Fig. 2.10. The performance curve labeled by "all" is to adopt the weighted sum of individual features as the new feature. The weight is chosen as the proportion to the fourth power of its individual accuracy.

From the results, we see that the correct classification rate of each individual feature ranges from 50–82 % while that of the "all" curve can perform up to 88.1 % in the scene recognition task for the fifteen scene dataset. However, when being applied to a much larger dataset, i.e. the SUN dataset, the performance of each individual feature drops to 5–28 % while the "all" curve drops to 38 %. This shows that, when the number of categories increases, the problem becomes more difficult.

In the work [14], we also see the differences between human and computer classification errors. Human errors mostly lie in semantically similar categories. However, computer errors are due to wrong features. The computer can make errors on semantically unrelated scenes. On the other hand, the computational methods are more accurate than human [14] for some categories.

A scene classification task is to classify scenes into different categories. A more difficult problem is scene detection. Being analogous with the object detection problem that identifies an sub-image as a certain object class, the scene detection problem is to identify a scene inside in an image. This problem arises from the observation

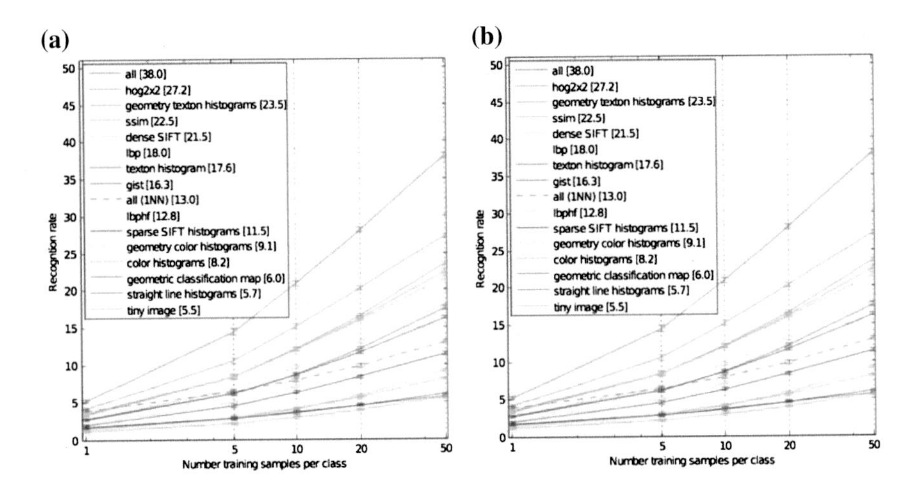

Fig. 2.10 Scene recognition with 24 scene types. **a** 15-scene dataset. **b** SUN dataset

that the real world scene might not be well divided into scenes of fifteen categories where each image exactly belongs to one scene. There might be several scene types in one image. To conduct the scene detection task, 24 well-sampled categories of the 398 categories are used in training. Another 104 photographs containing an average of 4 scene categories are used as test images. An example is shown in Fig. 2.11. Although only 24 scene categories are trained as possible scenes in the test image, the scene detection accuracy is not high, range from 8–66 % even using all features to train the classifier [14].

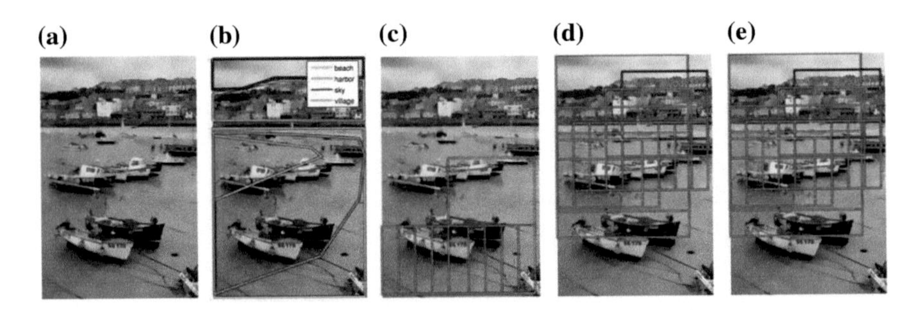

Fig. 2.11 **a** A photograph that contains multiple scene-types, **b** the ground truth annotations, **c–e** Detections of the beach, the harbor, and the village scene categories in one single image. In all visualizations, correct detections are in *green* and incorrect detections are in *red*. The bounding box size is proportional to classifier confidence. For this and other visualizations, all detections above a constant confidence threshold are shown. In this result, the harbor detection is incorrect since it does not overlap much with the ground truth "harbor" annotation while "beach" and "village" are acceptable

Both the scene classification performance comparison between human and computer and the performance of scene classification and detection using SVM show the difficulty of the scene understanding problem. A large scale dataset adds challenges to the problem. However, this database is needed to evaluate various algorithms and stimulate more advanced algorithms to be developed in the near future.

2.2.6 *Places205 Dataset*

With more and more data available and the emergence of deep learning trends, researchers stared to prepared even more larger dataset than SUN. Place205 [27] is the latest and most challenging one. Places205 contains 2,448,873 images from 205 scene categories in total. It is treated as the largest scene classification dataset, and mainly prepared for the purposes of Convolution Neural Network (CNN) training.

Comparing with ImageNet and SUN, in Fig. 2.12, Place205 shows extreme data abundance, which is very crucial for discriminative model learning for CNN with deep structures that has millions of parameters. In paper [27], author trained the huge CNN network with 2,448,873 image in 6 days and present superior results on traditional datasets with the trained deep features. In Fig. 2.13, we see the example output of Places-CNN for the query image on the left.

Since Places205 is proposed quite recently, and only CNN can take advantages of the such a large number of training images, there is little work benchmark using this dataset. However, it brought out a new definition of "large-scale image understanding".

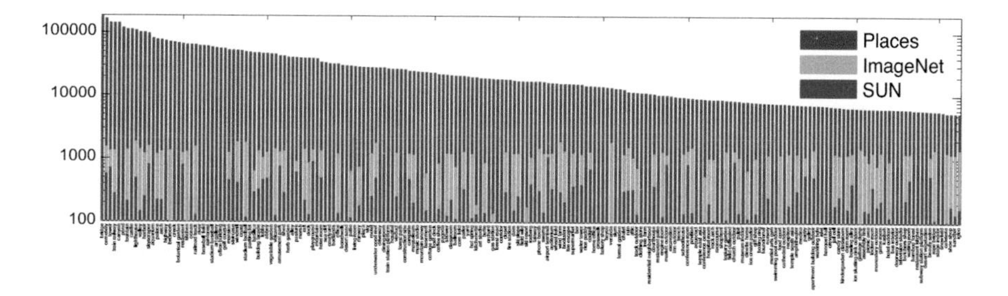

Fig. 2.12 Comparison of the numbers of images in Places 205 with ImageNet and SUN. Note that ImageNet only has 128 of the 205 categories, while SUN contains all of them. To compare them, we select a subset of Places. It contains the 88 common categories with ImageNet such that there are at least 1000 images in ImageNet. We call the corresponding subsets SUN 88 and ImageNet 88

(a)

(b)

Predictions

- **Type of environment:** indoor
- **Semantic categories:** restaurant:0.49, banquet_hall:0.14, restaurant_patio:0.13, cafeteria:0.08, coffee_shop:0.07
- **SUN scene attributes:** enclosedarea, electricindoorlighting, nohorizon, man-made, socializing, eating, working, congregating, cloth, wood(notpartofatree)
- **Informative region for the category "restaurant" is:**

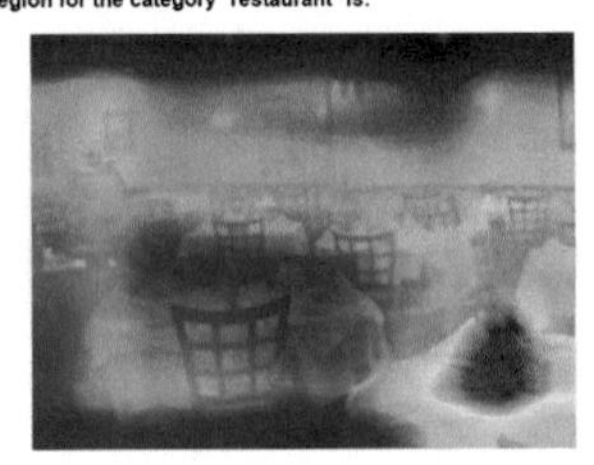

Fig. 2.13 Demo of the trained Places-CNN model: given a query image on the *left*, classification results are given with multiple classification soft scores

References

1. Oliva, A., Torralba, A.: Modeling the shape of the scene: a holistic representation of the spatial envelope. Int. J. Comput. Vision **42**(3), 145–175 (2001)
2. Liu, C., Yuen, J., Torralba, A.: Nonparametric scene parsing: label transfer via dense scene alignment. In: IEEE Conference on Computer Vision and Pattern Recognition, 2009. CVPR 2009, pp. 1972–1979. IEEE (2009)
3. Fei-Fei, L., Perona, P.: A bayesian hierarchical model for learning natural scene categories. In: IEEE Computer Society Conference on Computer Vision and Pattern Recognition, 2005. CVPR 2005, vol. 2, pp. 524–531. IEEE (2005)
4. Lazebnik, S., Schmid, C., Ponce, J.: Beyond bags of features: spatial pyramid matching for recognizing natural scene categories. In: IEEE Computer Society Conference on Computer Vision and Pattern Recognition, 2006, vol. 2, pp. 2169–2178. IEEE (2006)
5. Li, L.J., Fei-Fei, L.: What, where and who? classifying events by scene and object recognition. In: IEEE 11th International Conference on Computer Vision, 2007. ICCV 2007, pp. 1–8. IEEE (2007)
6. Hoiem, D., Efros, A., Hebert, M., et al.: Geometric context from a single image. In: Tenth IEEE International Conference on Computer Vision, 2005. ICCV 2005, vol. 1, pp. 654–661. IEEE (2005)
7. Hoiem, D., Efros, A., Hebert, M., et al.: Closing the loop in scene interpretation. In: IEEE Conference on Computer Vision and Pattern Recognition, 2008. CVPR 2008, pp. 1–8. IEEE (2008)
8. Torralba, A., Fergus, R., Freeman, W.T.: 80 million tiny images: a large data set for nonparametric object and scene recognition. IEEE Trans. Pattern Anal. Mach. Intell. **30**(11), 1958–1970 (2008)
9. Miller, G.A.: Wordnet: a lexical database for English. Commun. ACM **38**(11), 39–41 (1995)
10. Vogel, J., Schiele, B.: Natural scene retrieval based on a semantic modeling step (2004)
11. Everingham, M., Gool, L.V., Williams, C.K., Winn, J., Zisserman, A.: The pascal visual object classes (voc) challenge. Int. J. Comput. Vision **88**(2), 303–338 (2010)
12. Deng, J., Dong, W., Socher, R., Li, L.J., Li, K., Fei-Fei, L.: Imagenet: a large-scale hierarchical image database. In: IEEE Conference on Computer Vision and Pattern Recognition, 2009. CVPR 2009, pp. 248–255. IEEE (2009)
13. Russell, B.C., Torralba, A., Murphy, K.P., Freeman, W.T.: Labelme: a database and web-based tool for image annotation. Int. J. Comput. Vision **77**(1–3), 157–173 (2008)

14. Xiao, J., Hays, J., Ehinger, K.A., Oliva, A., Torralba, A.: Sun database: large-scale scene recognition from abbey to zoo. In: 2010 IEEE conference on Computer Vision and Pattern Recognition (CVPR), pp. 3485–3492. IEEE (2010)
15. Gao, T., Koller, D.: Discriminative learning of relaxed hierarchy for large-scale visual recognition. In: 2011 IEEE International Conference on Computer Vision (ICCV), pp. 2072–2079. IEEE (2011)
16. Pavlopoulou, C., Yu, S.X.: Indoor-outdoor classification with human accuracies: image or edge gist? In: 2010 IEEE Computer Society Conference on Computer Vision and Pattern Recognition Workshops (CVPRW), pp. 41–47. IEEE (2010)
17. Griffin, G., Holub, A., Perona, P.: Caltech-256 object category dataset (2007)
18. Dalal, N., Triggs, B.: Histograms of oriented gradients for human detection. In: IEEE Computer Society Conference on Computer Vision and Pattern Recognition, 2005. CVPR 2005, vol. 1, pp. 886–893. IEEE (2005)
19. Ojala, T., Pietiknen, M., Mnp, T.: Multiresolution gray-scale and rotation invariant texture classification with local binary patterns. IEEE Trans. Pattern Anal. Mach. Intell. **24**(7), 971–987 (2002)
20. Matas, J., Chum, O., Urban, M., Pajdla, T.: Robust wide-baseline stereo from maximally stable extremal regions. Image Vision Comput. **22**(10), 761–767 (2004)
21. Sivic, J., Zisserman, A.: Video data mining using configurations of viewpoint invariant regions. In: Proceedings of the 2004 IEEE Computer Society Conference on Computer Vision and Pattern Recognition, 2004. CVPR 2004, vol. 1, pp. I-488–I-495, IEEE (2004)
22. Hays, J., Efros, A.: Im2gps: estimating geographic information from a single image. In: IEEE Conference on Computer Vision and Pattern Recognition, 2008. CVPR 2008, pp. 1–8. IEEE (2008)
23. Zhang, J.K.W.: Video compass. Computer Vision? ECCV 2002, pp. 476–490. Springer (2002)
24. Martin, D., Fowlkes, C., Tal, D., Malik, J.: A database of human segmented natural images and its application to evaluating segmentation algorithms and measuring ecological statistics. In: Eighth IEEE International Conference on Computer Vision, 2001. ICCV 2001. Proceedings, vol. 2, pp. 416–423. IEEE (2001)
25. Hoiem, D., Efros, A.A., Hebert, M.: Recovering surface layout from an image. Int. J. Comput. Vision **75**(1), 151–172 (2007)
26. Lalonde, J.F., Hoiem, D., Efros, A.A., Rother, C., Winn, J., Criminisi, A.: Photo clip art. ACM Trans. Graph. **26**(3), 3 (2007)
27. Zhou, B., Lapedriza, A., Xiao, J., Torralba, A., Oliva, A.: Learning deep features for scene recognition using places database. In: Advances in Neural Information Processing Systems, pp. 487–495 (2014)

Chapter 3
Indoor/Outdoor Classification with Multiple Experts

Keywords Big visual data · Indoor/outdoor classification · Expert decision fusion · Structured machine learning system

3.1 Introduction

In the field of scene understanding, different experts (or classifiers) have different capabilities in handling different data types. The diversities among experts provides complementary values to each other. In this chapter, we propose a machine learning system to integrate multiple experts for the indoor/outdoor scene classification problem.

Indoor/outdoor scene classification is one of the basic scene classification problems in computer vision. Its solutions contribute to general scene classification [3, 10, 22, 31, 40, 46], image tagging [15, 33, 41], and many other applications [1, 2, 4, 48]. As compared to general scene classification problems, the indoor/outdoor scene classification problem has a clearer definition, namely, whether the scene is inside or outside a man-made structure with enclosed roofs and walls. Since the man-made structure is well-defined, the decision is unambiguous under various circumstances.

Indoor/outdoor classification allows a precise characterization of a wide range of images with diversified semantic meanings. For example, images from ① kitchen to ⑨ green house in the left column of Fig. 3.1 should all be classified as indoor images. In contrast with other scene classification problems [21, 40, 44], semantic objects in the scene may not help much in the decision. For example, indoor ⑤ swimming pool and outdoor ⑤ swimming pool in Fig. 3.1 share the same salient semantic object (i.e., the pool), yet they should be classified differently from the aspect of indoor/outdoor scene classification. The same observation occurs in quite a few real-world images.

Millions of images have been created every day thanks to the popularity of smart devices. Due to the huge sizes andgreat diversities of image data, applications

C. Chen et al., *Big Visual Data Analysis*,
SpringerBriefs in Signal Processing, DOI 10.1007/978-981-10-0631-9_3

Fig. 3.1 Exemplary indoor and outdoor scene images from the test dataset are given in the *left* and *right* of the *dash line*, respectively

such as large-scale image search [16] and tagging [23] will benefit from accurate indoor/outdoor classification results. Several methods, including SP [37], VFJ [42], SSL [34], PY [29], KPK [19] and XHE [47], were proposed to tackle this problem based on image datasets consisting of about 1,000 images. It is not clear whether the reported performance of these methods is scalable to large-scale datasets consisting of more than 100,000 images. This is one of the main focuses of our research in this chapter.

To address the large-scale indoor/outdoor scene classification problem, we propose an Expert Decision Fusion (EDF) system that consists of two key ideas—data grouping and decision stacking. In contrast with prior art, the proposed EDF system is less concerned with the search of new features but on a meaningful way to partition the dataset and organize basic indoor/outdoor classifiers in an effective way to lead to a more accurate and robust classification system. For convenience, each basic indoor/outdoor classifier is called an "expert" in this proposal.

The rest of this chapter is organized as follows. We describe several constituent experts in the EDF system in Sect. 3.2, which include existing indoor/outdoor scene classifiers as well as three newly developed classifiers. Then, the intuitions and methodologies of the EDF system is detailed. In Sect. 3.3, we show the performance gain from data grouping before training and testing using decisions from different experts. Then in Sect. 3.4, we present the diversity gain from combining multiple experts' decisions with stacking. Taking the advantage of data grouping and decisions stacking, in Sect. 3.5, we present the proposed structured system. Experimental results are reported and discussion is given in Sect. 3.6. Finally, concluding remarks and possible future extensions are presented in Sect. 3.7.

3.2 Individual Indoor/Outdoor Experts

3.2.1 Experts from Existing Work

SP

The SP expert [37], published in 1998, was a classical method for indoor/outdoor scene classification. It adopted a standard machine learning procedure and built a basic structure for model training and decisions making as shown in Fig. 3.2. Several later methods, such as SSL [34] and VFJ [42], followed the same structure with minor changes.

As shown in the flow chart of Fig. 3.2, basic operations of the SP method include: partitioning, feature extraction, model training and decision pooling. In the partitioning step, the original image is divided into 4×4 blocks for future feature extraction and model training. It is conducted so as to preserve a coarse spatial structure of the input image. In the feature extraction step, three low-level features (i.e., color, texture and frequency) are used as detailed below.

Ohta color feature. The Ohta color space [27] is used to extract local color features in SP. The axes of this color space are eigenvectors with the 3 largest eigenvalues in the RGB color space, which are determined by the principal components analysis of

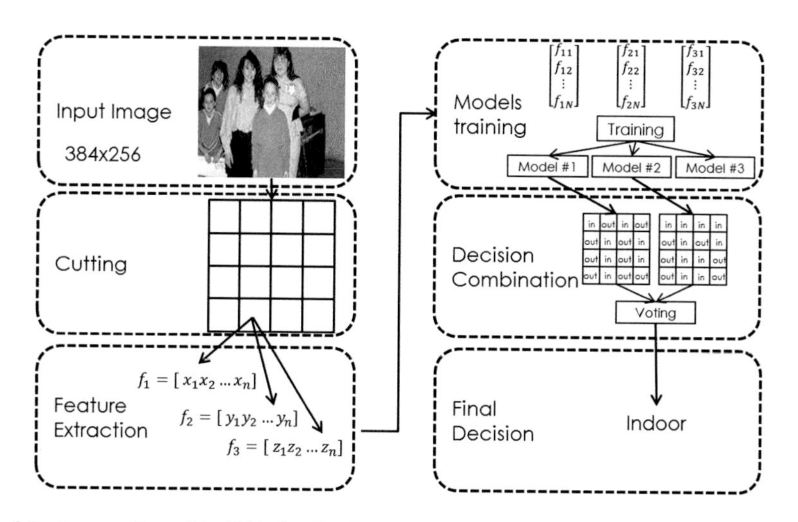

Fig. 3.2 An overview of the SP indoor/outdoor scene classification system (or expert)

a large selection of natural images. This yields the following:

$$\begin{aligned} I_1 &= R + G + B, \\ I_2 &= R - B, \\ I_3 &= R - 2G + B, \end{aligned} \tag{3.1}$$

where I_1, I_2 and I_3 are 3 channels in the Ohta color space, respectively. The Ohta color channels are approximately decorrelated so that they offer a good choice for per-channel histogram computation. The change from the RGB color space to the Ohta color space raised the performance of the color-histogram-based classification from 69.75 to 73.2 % in the experiment reported in [37]. To extract the color features of a block, SP computed a 32-bin uniform histogram along each Ohta channel, and concatenated the histograms of three channels to make a 96-dimensional color feature vector. To calculate the distance between two histograms, the histogram intersection norm [36] was used. It measures the amount of overlap between the corresponding buckets in two histograms h^1 and h^2 as given by

$$dist(h^1, h^2) = \sum_{i=1}^{N} (h_i^1 - min(h_i^1, h_i^2)). \tag{3.2}$$

When both the Ohta color space and the histogram intersection methods are used, the classification rate increases from 69.75 to 74.2 %.

MSAR texture feature. The multi-resolution simultaneous autoregressive model (MSAR) [25] was used as the texture feature in SP. It was one of popular texture features in early days when benchmarked on the Brodatz album [30]. The MSAR offers a good linear predictor of a pixel based on its non-causal neighborhood. In the implementation, the SP method constructed a 15-dimensional feature vector obtained from three scales for each block and used the Mahalanobis norm to measure the distance between these vectors.

DCT. The 2-D Discrete Cosine Transform (DCT) is used in SP as frequency features. In the implementations, sub-blocks of size 8×8 are used in the DCT computation, and results are averaged over all sub-blocks in each block. Since periodic textures will show a regular pattern of peaks (fundamental and harmonic frequencies) after the DCT, coefficients of all corresponding frequencies can be averaged to yield one single coefficient. Then, the Mahalanobis distance metric can be used to compute the vector distances for the DCT feature vector.

When training a model for each feature type (color, texture and frequency), SP collected the feature vector of all blocks in all training image, and used the K-Nearest Neighbor (K-NN) classification tool to train models for this feature type. In other words, each feature type will have its own model and a sample in the training process is the feature vector of a single block of an input image.

To get the final decision for an input image, SP only uses the trained color and texture models to get color-model decision and texture-model decision for blocks.

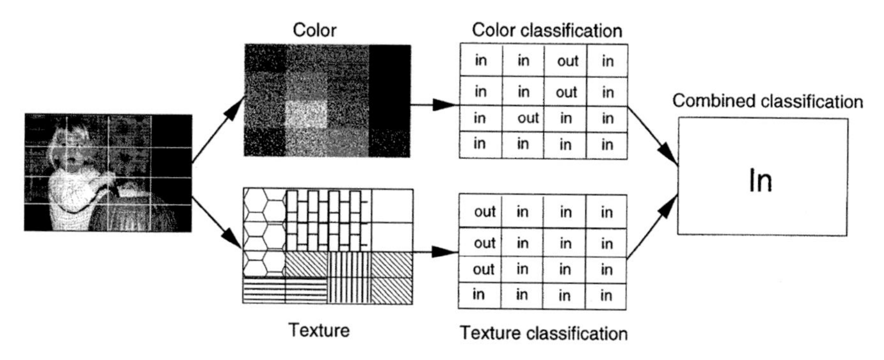

Fig. 3.3 Decision fusion via majority voting in SP

As a result, each block has two decisions—one from the color model and the other from the texture model. As shown in Fig. 3.3, SP obtains 32 block decisions for each image. Afterwards, a majority voting rule was adopted to get the final decision of the input image.

The performance of three experimental setups for the Kodak consumer photo dataset was reported in [37]. Table 3.1 shows the K-NN classification performance on blocks with different K values. As K increases, the performance improves and reaches a saturation level of 74.7 %. With the help of majority voting, the performance of each individual feature increases significantly as shown in Table 3.2, where the best performance can go up to 85 % with $K = 5$. By combining the Ohta color feature and the MSAR texture feature, SP can reach a classification accuracy of 90.3 % as shown in Table 3.3.

The framework of the SP expert dominated the scene classification research for quite some time, where multiple features were evaluated theoretically and experimentally. Besides, simple decisions based on different features were integrated by majority voting. Although it was tested on a small dataset and its performance can be further improved, it still provides a valuable reference for our consideration today.

VFJ

The VFJ expert [42] was proposed shortly after SP in 2001, where indoor/outdoor classification was treated as the first-stage classification task in a hierarchical

Table 3.1 Performance of block-based K-NN classification with different K values

Feature	K = 1	K = 3	K = 5	K = 9	K = 13
Ohta histogram intersection	66.0	68.1	69.2	70.1	70.3
MSAR half resol	69.2	72.0	73.1	74.2	74.7
MSAR quarter resol	64.5	67.6	68.9	70.0	70.6
DCT half resol	66.4	69.7	71.0	72.0	72.3

Table 3.2 Performance of fused decisions from block-based K-NN classification via majority voting with a single feature

Feature	K = 1	K = 3	K = 5	K = 9	K = 13
Ohta histogram intersection	78.2	80.2	81.0	81.0	80.5
MSAR half resol	82.0	84.4	85.0	84.5	84.0
MSAR quarter resol	80.0	83.0	81.9	82.6	84.0
DCT half resol	82.0	84.4	85.0	84.5	84.0

Table 3.3 Performance of fused decisions from block-based K-NN classification via majority voting with multiple features

Feature	Performance
Color, MSAR	90.3
Color, DCT	89.0
MSAR, DCT	86.5
Color, MSAR, DCT	89.9

content-based indexing system as illustrated in Fig. 3.4. On one hand, since the performance of the first-stage classification has a huge impact on the performance of the whole system, an accurate and robust indoor/outdoor classification solution is in demand. On the other hand, the main focus of [42] was the proposal of an integrated indexing system, discussion on VFJ is brief. The flowchart of VFJ can be seen in Fig. 3.5.

Being similar to SP, ideas such as image partition, feature extraction, model training were used as shown in Fig. 3.5. In implementation details, VFJ partitions an image into $100 (= 10 \times 10)$ blocks instead of $16 (= 4 \times 4)$ blocks. The objective is to increase the geometric resolution of extracted local features in an image. Although both color and texture features were mentioned in [42], only the color feature was implemented in the experiments. The color moments of the LUV color channels of a block are used as its color feature. In the final step, VFJ combines color features of each block by concatenating them to a single feature vector.

LUV color moments. The LUV color space is adopted by VFJ to evaluate the color property of an image. Typically, the RGB color space is first transformed to the XYZ color space and, then, the XYZ color space is transformed to the LUV space. According to theoretical analysis and experimental evidences in [40], LUV moments yield better results in image retrieval than moments in other spaces. The VFJ expert uses the first-order and second-order moments of the three channels of the LUV space. Since each image is divided into 100 blocks and there are 6 color moment features in each block, the feature vector of an image has a dimension of 600.

Bayesian Classifier and Vector Quantization.

Bayesian methods have been successfully applied to many image analysis tasks [43]. In image classification, the decision can be made based on the distribution of m feature sets denoted by,

$$\mathbf{y}^{(1)}, \mathbf{y}^{(2)}, \ldots, \mathbf{y}^{(m)}, \tag{3.3}$$

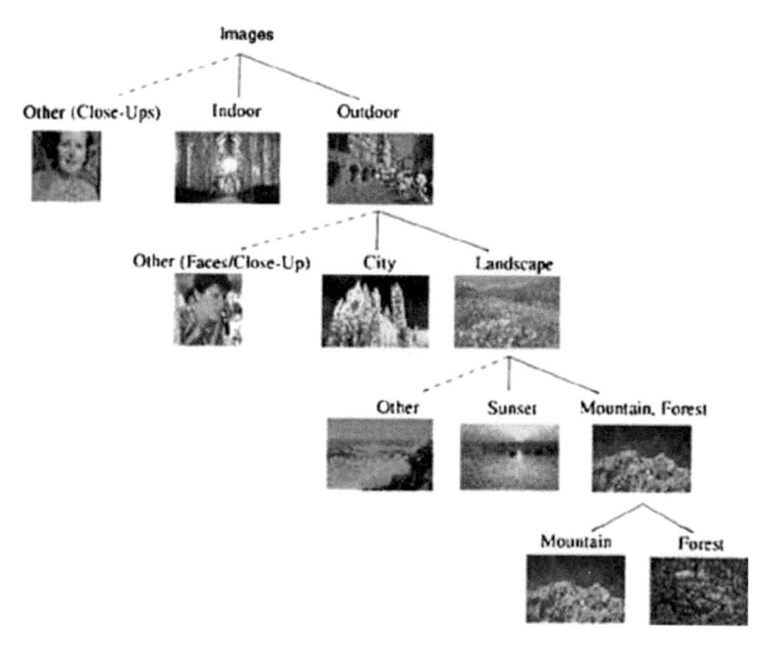

Fig. 3.4 The hierarchy of semantic image classification where the problems addressed by VFJ were indicated by *solid lines*

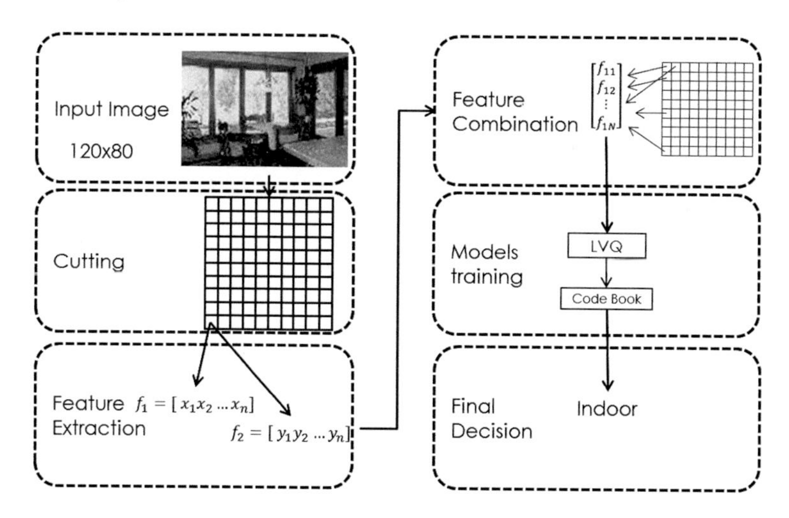

Fig. 3.5 An overview of the VFJ indoor/outdoor scene classification system (or expert)

which are class-conditionally independent. Mathematically, we have

$$f(\mathbf{y}|w) = \prod_{i=1}^{m} f(\mathbf{y}^{(i)}|w), \quad \text{for } w \in \Omega \tag{3.4}$$

where Ω is the set of classes. For the case of indoor/outdoor scene classification, there are two classes. To infer the proper class type based on the Maximum A Posteriori (MAP) criterion, we have

$$\hat{w} = \arg\max_{w \in \Omega} p(w|\mathbf{y}) = \arg\max_{w \in \Omega} f(\mathbf{y}|w)p(w). \tag{3.5}$$

To estimate class-conditional densities $f(\mathbf{y}|w)$, VFJ adopts the vector quantization technique [13] by leveraging an open source package called LVQ_PAK [20]. Furthermore, VFJ used a modified MDL [32] scheme to select the optimal codebook size.

In Fig. 3.6, we show the performance of the VFJ indoor/outdoor classifier as a function of the codebook size. The classifier is trained on 2541 images and tested on an independent set of 2540 images. We see two performance peaks when the codebook size is equal to 15 and 25, respectively, in Fig. 3.6. VFJ uses a codebook of size 15. The performance of two disjoint data sets, called Test Set 1 and Test Set 2, is shown in Table 3.4, where VFJ offers an acceptable performance. This is possible due to the limited size and diversity of the entire dataset.

To conclude, VFJ uses only the color moments in the LUV space as the feature vector for indoor/outdoor classification. As compared to the dataset reported in the SP work, a larger dataset with around 7,000 images is used in VFJ's experiments with a correct classification rate of 88 %. However, the dataset used in [42] is not available to the public, and one can only have a glimpse of a small part of their Corel dataset.

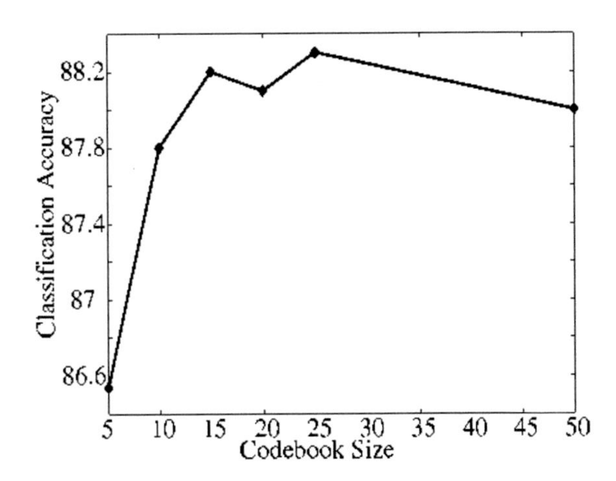

Fig. 3.6 The performance of the VFJ indoor/outdoor classifier as a function of the codebook size, which is trained on 2541 images and tested on an independent set of 2540 images

Table 3.4 The performance of the VFJ indoor/outdoor classification expert as reported in [42]

Test Data	Database size	Accuracy (%)
Training Set	2541	94.2
Test Set 1	2540	88.2
Test Set 2	1850	88.7
Entire Database	6931	90.5

SSL

The SSL expert [34] is very similar to SP with only differences in feature selection and replacement of majority voting with the SVM classifier. The structure of the SSL system is shown in Fig. 3.7. The SSL system offers more efficient color and texture features than the Ohta color and the MSAR texture feature adopted by SP. Furthermore, the SVM tool takes the soft decision (i.e. the distance to classification boundaries) of each block as the input and and provides more robust and accurate classification performance.

LST color feature. The LST color space is chosen as the color feature in SSL. The purpose is to decorrelate the original RGB color channels. Actually, the LST color space is akin to the Ohta color space used in SP. The RGB space can be transformed to the LST space via

$$L = \frac{k}{\sqrt{3}}(R+G+B), \quad S = \frac{k}{\sqrt{2}}(R-B), \quad T = \frac{k}{\sqrt{6}}(R-2G+B), \tag{3.6}$$

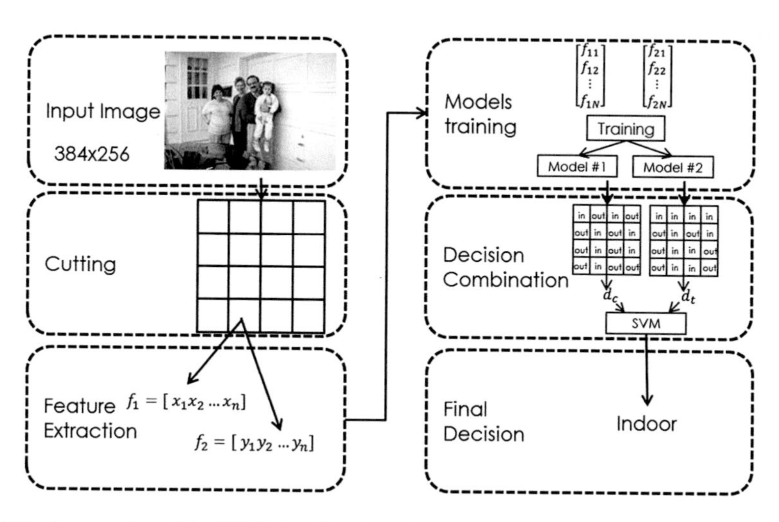

Fig. 3.7 An overview of the SSL indoor/outdoor classification system (or expert)

where L is the luminance channel, S and T are two chrominance channels, and $k = \frac{255}{\max R,G,B}$. By comparing Eqs. 3.1 and 3.6, we see that the LST color space is a normalized version of the Ohta color space. Instead of the 32-bin histogram in each channel used in SP, SSL uses the 16-bin histogram to reduce the influence of noise. A concatenation of three histograms leads to a 48-D color feature.

Wavelet texture feature. In SSL, the texture feature is obtained from a two-level wavelet decomposition. The decomposition is performed on the L-channel using Daubechies' 4-tap filters [7], where the low-pass filter, $h(n)$, and the high-pass filter, $g(n)$, are of the following forms:

$$h(n) = [-0.1290.2240.8370.483], \quad g(n) = [-0.4830.837 - 0.224 - 0.129]. \tag{3.7}$$

They are applied to both rows and columns of an image separately with down-sampling to yield a 2-D separable wavelet transform as shown in Fig. 3.8, where $c_2, c_3, c_4, c_5, c_6, c_7$ and c_8 represent sub-band coefficients. The low-frequency coefficient c_5 is replaced by the Laplacian filtered one [24] since the texture has little low-frequency component. Then, texture features can be obtained by computing the energy function of the following seven sub-bands:

$$e_k = \frac{1}{M_k N_k} \sum_{i=1}^{M_k} \sum_{j=1}^{N_k} |c_k(i,j)|^2, \quad k = 2, 3, \ldots 8, \tag{3.8}$$

where M_k and N_k are dimensions of subband k. SSL uses wavelets to increase computational efficiency and reduce the dimension of the texture feature vector. In contrast with the 15-D MSAR features used in SP, its texture dimension is only seven.

Support Vector Machine (SVM) and Decision Pooling. The support vector machine (SVM) is a supervised learning tool that analyzes data for classification and regression. The original SVM algorithm was originally proposed by Vapnik, and its current form (soft margin) was developed by Cortes and Vapnik [6]. It aims at maximum margin classification and reduces the error caused by traditional linear or nonlinear classification boundaries. It was first adopted by SSL for the indoor/outdoor classification problem.

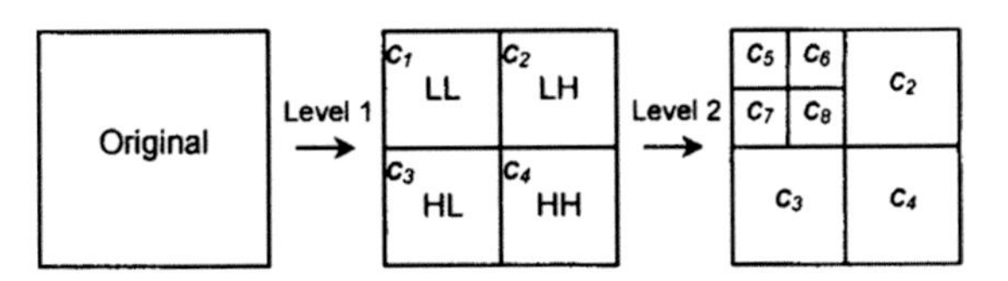

Fig. 3.8 Illustration of the two-level pyramid wavelet transform

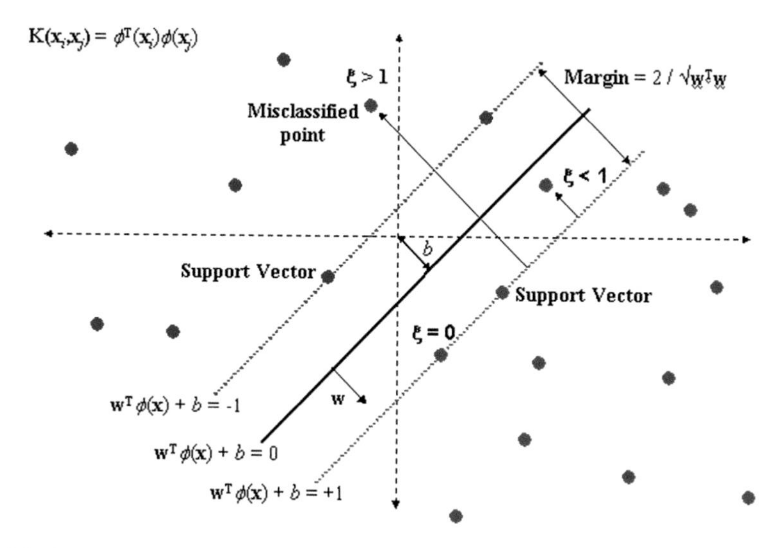

Fig. 3.9 Illustration of the maximum-margin hyper-plane and its margins to trained samples from two classes, where the samples lying on the margin are called support vectors

To achieve the optimal separating hyper-plane in Fig. 3.9, we can solve for the following equation [5]:

$$\min_{w,b,\xi} \frac{1}{2} w^t w + C \sum_{i=1}^{l} \xi_i, \quad i = 1, \ldots, l, \tag{3.9}$$

subject to $y_i(w^T \phi(x_i) + b) \leq 1 - \xi_i$ and $\xi_i \leq 0$, where w is the model coefficients, $C > 0$ is the regularization parameter, ξ is the soft-margin error function, ϕ is the linear or nonlinear mapping of features and y is the ground truth. This is called the prime form of the SVM objective function. By computing parameters w and b, we can obtain the maximum-margin hyper-plane and use it as the SVM classifier.

SSL applies the SVM classification tool to the indoor/outdoor classification problem to replace the K-NN and voting schemes in SP. Finally, a block-based SVM classifier and a decision combination SVM classifier are trained in the whole process. For the latter, SSL first summarizes soft decisions of all blocks using the following rule:

$$d_c = \sum_{i=1}^{16} f_c(x_c^i), \quad d_t = \sum_{i=1}^{16} f_t(x_t^i), \tag{3.10}$$

where d_c and d_t are global color and texture features and $f_c(x_c^i)$ and $f_t(x_t^i)$ are collected sample-to-hyperplane distances for all training images, respectively. Afterwards, the 2-D feature vectors will be combined.

Table 3.5 The block classification performance of the SSL expert

Feature	Training set (%)	Test set (%)
Color	73.7	67.6
Texture	75.8	73.0

Table 3.6 The final classification performance of the SSL expert

Classifier	Training set (%)	Test set (%)
Majority classifier	92.8	87.2
Second stage SVM	95.0	90.2

The block classification rate of SSL using the same dataset of SP is shown in Table 3.5. Its poor performance is attributed to a smaller block size. On the other hand, there is a significant improvement after the application of the SVM classifier in decision combination of 2-D feature vectors of all blocks, which is shown in Table 3.6. It demonstrates the power of the SVM meta-classifier in decision fusion.

To conclude, the SSL expert uses the LST histogram color feature and the wavelet texture feature. With a more powerful SVM classifier in fusing block decisions, SSL improves the performance of SP on the same dataset, which contains only about 1,300 images. Overfitting and performance robustness could be serious problems for such a small dataset. Actually, the SSL expert does not offer a competitive performance in our experiment, which is tested on a much larger database consisting of more than 100,000 images.

PY

The PY method [29] exploits a powerful global scene descriptor, GIST, to boost the indoor/outdoor scene classification performance. GIST [28] is a global multi-scale texture analysis feature. It has been widely applied to the scene classification problem [21, 26]. The system diagram of the PY expert is shown in Fig. 3.10.

The computation of the GIST feature is illstrated in Fig. 3.11. The input is usually in a gray scale image since GIST is a local structure pattern descriptor that has little to do with the color information. The input image is divided into $16(= 4 \times 4)$ blocks. Then, the Gabor filter [11] is used to extract the local texture pattern, where the Gabor filter is a 2-D Gaussian kernel function modulated by a 2-D sinusoidal wave. It can be expressed as

$$g(x, y, \lambda, \theta, \psi, \delta, \gamma) = exp(-\frac{x'^2 + \gamma^2 y'^2}{2\delta^2})exp(i(2\pi\frac{x'}{\lambda} + \psi)), \tag{3.11}$$

where λ is the wavelength of the sinusoidal factor, θ represents the orientation of the normal to parallel stripes of a Gabor function, ψ is the phase offset, δ is the sigma/standard deviation of the Gaussian envelope, and γ is the spatial aspect ratio

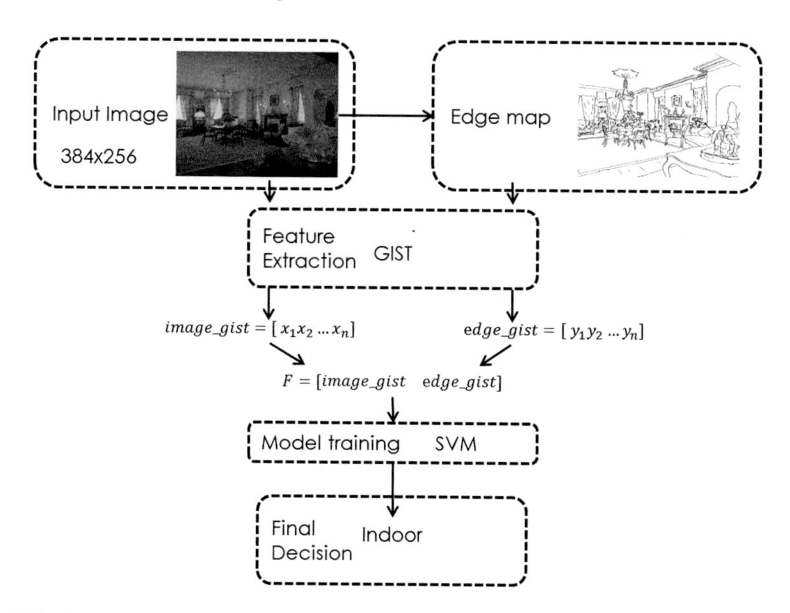

Fig. 3.10 An overview of the PY indoor/outdoor classification system (or expert)

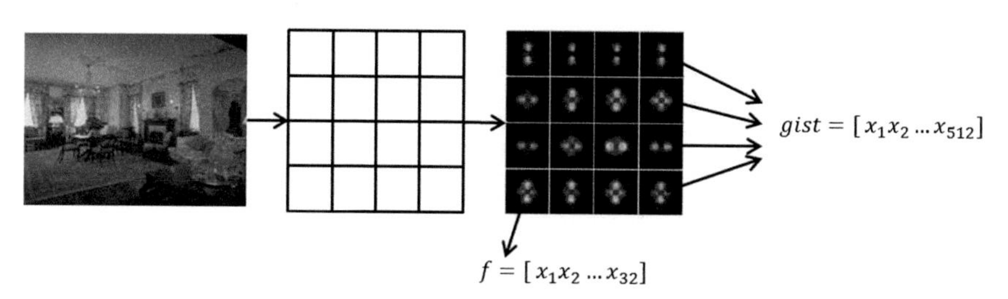

Fig. 3.11 Computation of the GIST feature

that specifies the ellipticity of the support of the Gabor function. Typically, a GIST descriptor contains 4 wavelength scales and 8 orientations. For every orientation in each scale, the averaged energy of the filter's response is calculated as one feature. Thus, each block has a 32-D feature vector. For the whole image, we concatenate the 32-D feature vectors of 16 blocks to result in a $512 (= 32 \times 16)$ dimensional GIST descriptor. Furthermore, the PY method extracts the GIST descriptor from the gray-scale image as well as its edge map. In other words, we obtain the image GIST feature and the edge GIST feature separately. Finally, each of them is fed into an SVM classifier for training or testing.

The dataset used to test the performance of the PY method is larger. It consists of around 20,000 images from multiple sources, including [31, 33, 35], and Flickr, yet it is not released to the public. The performance of the PY method using the image and edge GIST features is shown in Table 3.7. We see that the edge GIST

Table 3.7 Performance comparison of the PY expert using the image GIST and the edge GIST features

	Indoor (%)	Outdoor (%)
Image GIST	57	76
Edge GIST	66	82

feature outperforms the the image GIST by a significant margin for the indoor class. Indoor scenes often consist of complex illumination conditions, e.g., natural lighting coming through windows, artificial lighting from various sources, multiple surfaces with different reflection properties. The edge GIST is more robust to these changes.

To conclude, PY takes advantage of two advanced tools, GIST and SVM. Besides, PY applies GIST to edge maps and achieves better results on both indoor and outdoor cases. Although the reported performance is not as high as SP, PS and SSL, its database is larger so that such a direct comparison is not meaningful. It turns out that the performance of PY in our experiments on the SUN dataset is closest to the numbers claimed in [29] while the performance of others drops significantly.

KPK

The block-diagram of the KPK indoor/outdoor classification system [19] is shown in Fig. 3.12. As shown in the figure, KPK follows a similar framework in image training and decision-making. It has three distinctive characteristics: non-uniform image partition, new features, and weighted feature combination.

Non-uniform Image Partitioning. A non-uniform image partition scheme is adopted by KPK as shown in Fig. 3.13. This partition is inspired from the observation that most images contain objects of interest in the central regions, yet these objects rarely play a role in indoor/outdoor image classification. Thus, instead of uniform partitioning,

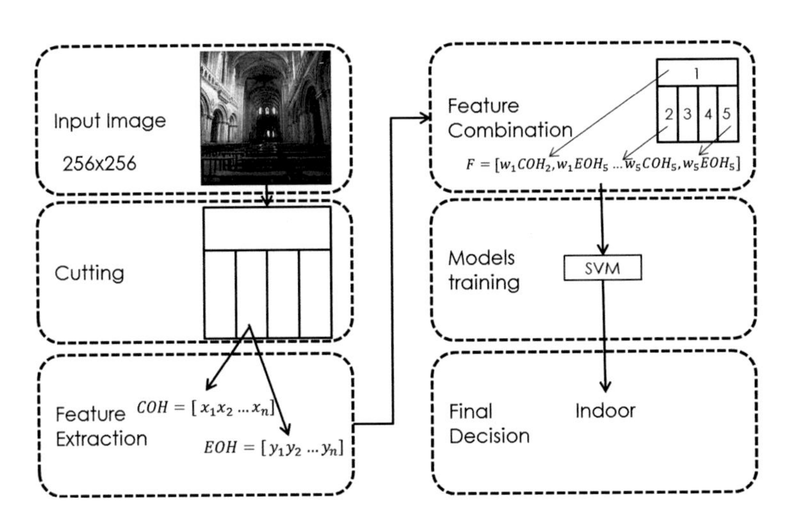

Fig. 3.12 An overview of the KPK indoor/outdoor classification system

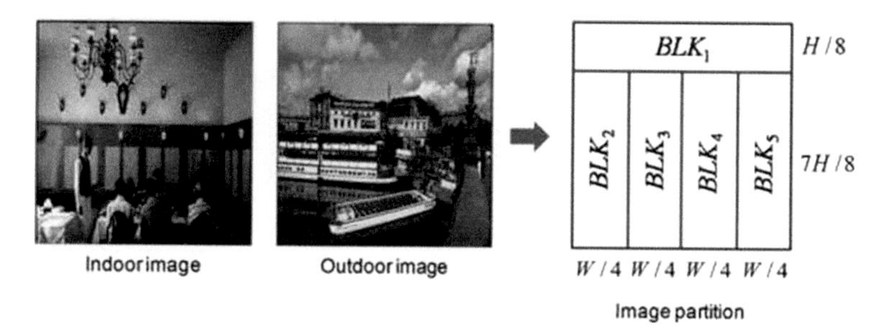

Fig. 3.13 The image partition scheme adopted by KPK

KPK partitions an image into to two central blocks (BLK_3 and BLK_4) and three boundary blocks (BLK_1, BLK_2 and BLK_5).

Color Orientation Histogram (COH) and Edge Orientation Histogram (EOH).

The COH is computed using the HSV color model [12] as illustrated in Fig. 3.14. The procedure to transform from the RGB color space to the HSV color space can be written as:

The values of R, G, B three color channels are all normalized to [0,1].

$$
\begin{aligned}
C_{max} &= \max(R, G, B) \\
C_{min} &= \min(R, G, B) \\
\Delta &= C_{max} - C_{min} \\
H &= \begin{cases} 60 \times (\frac{G-B}{\Delta} \mod 6) & C_{max} = R \\ 60 \times (\frac{B-R}{\Delta} + 2) & C_{max} = G \\ 60 \times (\frac{R-G}{\Delta} + 4) & C_{max} = B \end{cases} \\
S &= \begin{cases} 0 & \Delta = 0 \\ \frac{\Delta}{C_{max}} & \Delta \neq 0 \end{cases} \\
V &= C_{max}
\end{aligned}
\tag{3.12}
$$

Fig. 3.14 The HSV color space

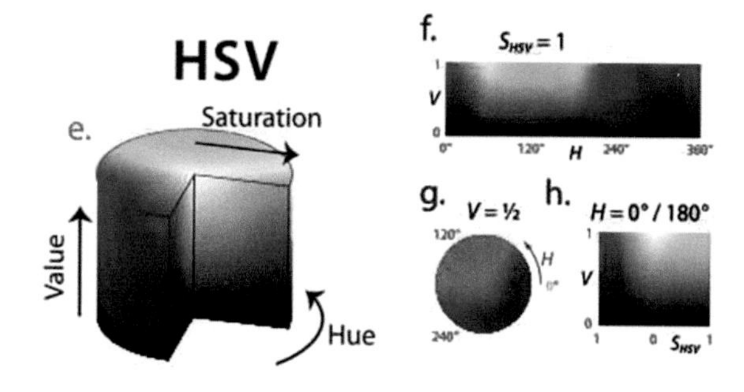

The hue value represents a color tone with an angle in the HSV color space. The values are quantized into K color orientation histogram bins. The parameter, K, has an impact on the classification result, and KPK chooses an empirical K value to optimize the performance. Purer colors have a larger saturation value. Moreover, different illumination conditions between indoor and outdoor environments yields different saturation values. For these reasons, KPK generates the COH feature by accumulating the saturation value to the corresponding hue value bin via

$$F_{i,n}^{COH} = \frac{C_{i,n}}{\sqrt{\sum_{j=1}^{K}(C_{i,j})^2 + \varepsilon}}, \quad 1 \le i \le 5, \text{ and } n = 1, \ldots, K, \tag{3.13}$$

where

$$C_{i,n} = \sum s(x, y), \quad (x, y) \in BLK_i, \quad h(x, y) \in n, \tag{3.14}$$

and where $s(x, y)$ and $h(x, y)$ are the saturation value and the quantized hue value at pixel position (x, y), respectively. Thus, rigorously speaking, the so-called color orientation histogram is actually the hue-saturation channel histogram.

A procedure similar to COH is used to to extract the edge orientation histogram (EOH). The edge orientation is first calculated and quantized into K angle bins. Next, the K-bin histogram is generated by accumulating the edge magnitude in each edge orientation bin over pixels in a block. Mathematically, the K-bin edge orientation histogram is given by

$$F_{i,n}^{EOH} = \frac{E_{i,n}}{\sqrt{\sum_{j=1}^{K}(E_{i,j})^2 + \varepsilon}}, \quad 1 \le i \le 5, \tag{3.15}$$

where $n (= 1, \ldots, K)$ is the histogram index, ε is a small positive constant to ensure robustness if the whole region is edge-free, and

$$E_{i,n} = \sum m(x, y), \quad (x, y) \in BLK_i, \quad \theta(x, y) \in n, \tag{3.16}$$

and where $m(x, y)$ and $\theta(x, y)$ are the edge magnitude and the quantized orientation at pixel position (x, y), respectively.

After extracting COH and EOH feature vectors from all five blocks, we concatenate them into one single long feature vector:

$$F = w_1 F_1^{ECOH}, w_2 F_2^{ECOH}, \ldots, w_5 F_5^{ECOH}, \tag{3.17}$$

where w_i is the weigh for the ith block and

$$F_i^{ECOH} = (F_{i1}^{EOH}, F_{i1}^{COH}, F_{i2}^{EOH}, F_{i2}^{COH}, \ldots, F_{iK}^{EOH}, F_{iK}^{COH}). \tag{3.18}$$

In above, weights w_i should be properly determined to achieve higher classification performance. It can be determined by the confidence value of each block as detailed below. The confidence value of a block is calculated based on the training data. The averaged confidence value for the ith block is given by

$$d_i = \frac{1}{N_{in}} \sum_{j=1}^{N_{in}} f^i(x_j^{in}) + \frac{1}{N_{out}} \sum_{j=1}^{N_{out}} |f^i(x_j^{out})|, \tag{3.19}$$

where N_{in} and N_{out} denote the number of indoor and outdoor training images, respectively, and $f^i(x_j^{in})$ and $f^i(x_j^{out})$ are the confidence values for the jth sample of the ith block in the indoor and outdoor dataset. Note that the absolute confidence value is used in the second term of the right-hand-side of Eq. (3.19) since the confidence value becomes negative for outdoor images. Finally, the weight value is the normalized sum of the average confidence value as

$$w_T = \sum_{i \in T} \frac{d_i}{\sum_{j=1}^{5} d_j}, \tag{3.20}$$

where T denotes either the boundary region (BLK1, BLK2 and BLK5) or the center region (BLK3 and BLK4). In practice, the weights for the boundary region and the central region are simply set to $w_{br} = 0.67$ and $w_{cr} = 0.33$ by KPK in the implementation.

The quantization parameter, K, is selected based on an experiment using 600 indoor images and 600 outdoor images. The classification performance is plotted as a function of four K values in Fig. 3.15. Since $K = 8$ provides the best performance among the four choices, KPK selects 8 quantization levels in the implementation. Consequentially, the combined ECOH feature has a dimension of 80 for the whole image (16 for each of five blocks).

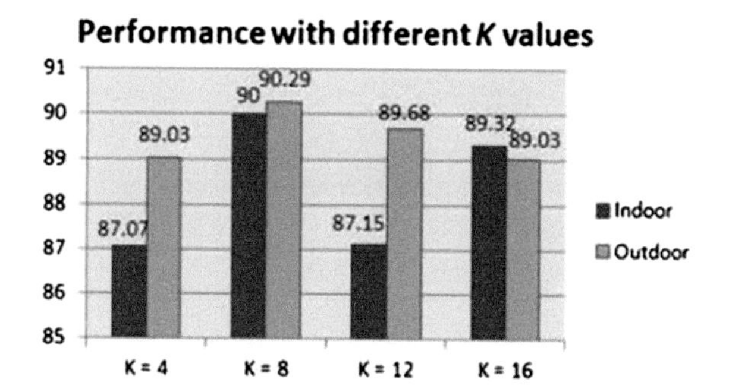

Fig. 3.15 Performance comparison of the KPK indoor/outdoor classification system with respect to four quantization levels (K = 4, 8, 12 and 16)

Table 3.8 Reported performance of KPK

Descriptors	Dimension	Indoor set (%)	Outdoor set (%)	Total set (%)
EOH	40	70.65	80.51	77.00
COH	40	85.81	83.91	84.82
Ohta	96	66.61	64.84	68.35
ECOH	80	90.00	90.29	90.26

The overall KPK classification performance is 90.26 %, with 90.00 % on the indoor class and 90.29 % on the outdoor class. Comparison with the Ohta color feature used by SP is given in Table 3.8.

To summarize, the KPK expert adopts a non-uniform partitioning scheme and puts more weights on the boundary blocks than the central ones. The COH and EOH features are carefully selected. In our experiment on the SUN dataset, KPK provides the best performance and the observed performance is close to that claimed in [19] despite the fact that only a small dataset consisting of 1,200 images was used therein.

XHE

XHE [47] is one of the features proposed in the study of the SUN database, which is built for modern scene classification related research with a large number of scene images of diversified content. More about this dataset will be discussed in Chap. 4. Here, we choose XHE to be an indoor/outdoor classification expert whose system diagram is shown in Fig. 3.16.

First, XHE down-samples an input image into a tiny image of size 16×16 since it is argued in [18, 39] that a tiny image can maintain the global color structure of an image and provide cues for human to recognize the input image category. Thus, XHE concatenates the RGB values of pixels to make a 768-D feature vector, which is the global color distribution descriptor of the input image. Finally, a SVM classifier is trained and used as the indoor/outdoor classification model.

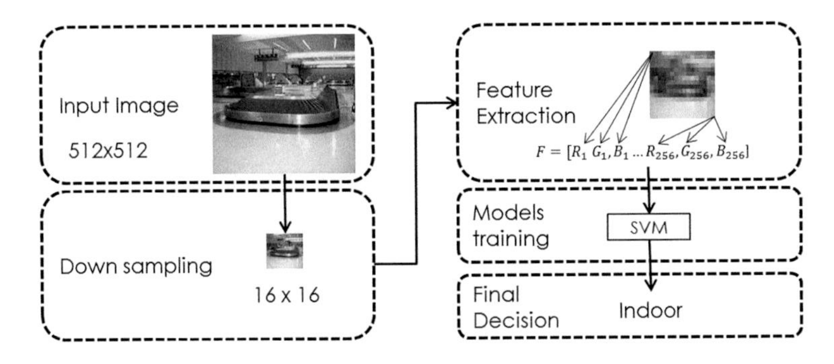

Fig. 3.16 An overview of the XHE indoor/outdoor classification system

It is worthwhile to point out that XHE was designed to discriminate general scene categories and no specific experiment was conducted to evaluate XHE's capability on indoor/outdoor classification [47]. Since XHE is relatively simple as compared to other scene classification experts, we do not expect high classification performance. On the other hand, XHE does provide complementary strength to other methods so that the performance of the decision fusion system can be improved furthermore.

3.2.2 Proposed Experts

HSH

The block diagram of the HSH classification system in shown in Fig. 3.17. As shown in the figure, we do not partition an input image into blocks since we attempt to evaluate the distribution and correlations of the Hue and Saturation color channels for the entire image. An 80-bin global hue-saturation histogram is used as the feature vector and the standard SVM training process is used to obtain an indoor/outdoor classifier.

Hue-Saturation Histogram. Instead of considering the color histogram in a block as adopted by KPK, HSH examines the HSV color distribution in a full image. The light sources of indoor and outdoor scenes are quite different. For indoor scenes, weak and close point-light sources (human made light sources such as light bulbs) with reflections from plain surfaces such as tables, floors and walls make colors of objects purer (i.e., with a higher saturation value). On the other hand, for outdoor scenes, natural light sources are stronger and parallel in direction and objects tend to be pale in color (i.e., with a lower saturation value). Two examples are shown in Fig. 3.18, where the bedroom scene has more saturated colors than the beach scene. Thus, we can use the hue-saturation histogram to discriminate indoor and outdoor scenes.

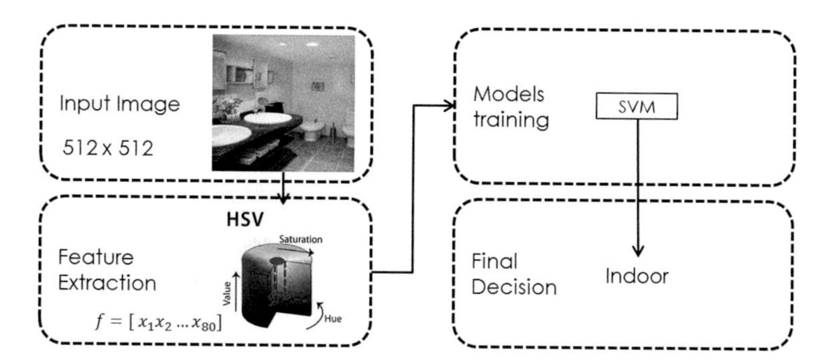

Fig. 3.17 An overview of the HSH indoor/outdoor classification system

Fig. 3.18 Visual comparison of hue and saturation differences in two exemplary indoor and outdoor scenes. **a** Bedroom. **b** Beach

To extract the hue-saturation histogram of an image, we first define a proper volume in the HSV color space. We observe that colors that have low V and S channel values do not contribute much to discrimination of indoor/outdoor scenes. Let h, s, and v denote the HSV value of a pixel. If $v > T_v$ and $s > T_s$, where T_v and T_s are two threshold values, we count the pixel in the histogram computation. Otherwise, it is discarded. In the implementation, we set $T_v = 0.2$ and $T_s = 0.1$ empirically. This idea is illustrated in Fig. 3.19. Then, we quantize the hue values into 16 bins. Each hue bin is further divided into 5 saturation bins. Consequently, it leads to a 80-bin hue-saturation histogram, which is used as the feature vector for training and testing.

The HSH expert exploits the global hue-saturation distribution for indoor/outdoor classification. It is relatively simple since there is no image partition and decision fusion in the system. It will serve as a supplement to other indoor/outdoor classification experts.

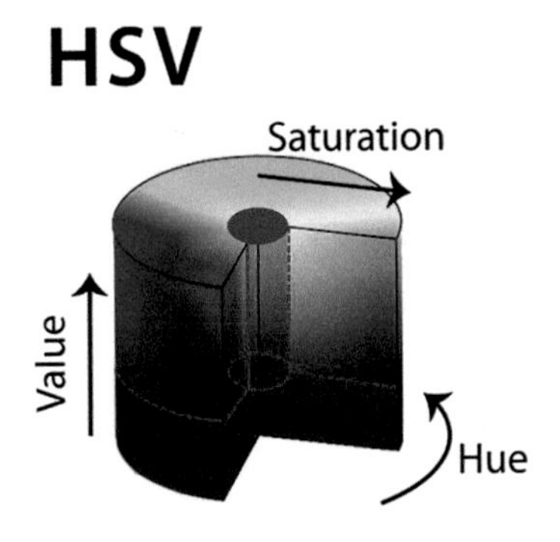

Fig. 3.19 The HSH expert selects a specific region in the HSV space for the hue-saturation histogram computation. *Red color* masked region in the figure is not used in the HSH

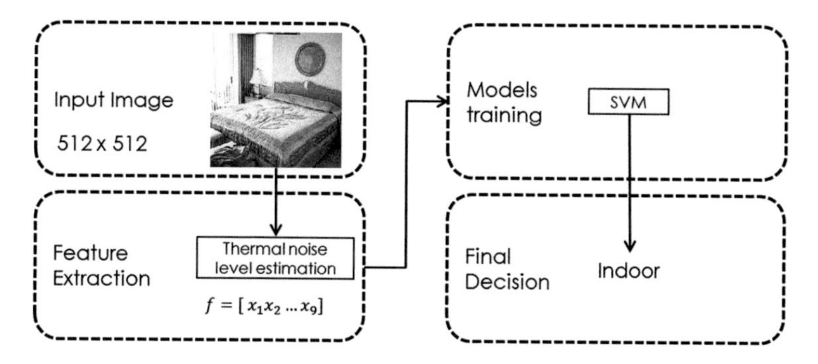

Fig. 3.20 An overview of the TN indoor/outdoor classification system

TN

The acronym, TN, stands for the "Thermal Noise". The TN expert classifies scene images by their digital noise levels. With different lighting sources and properties, indoor and outdoor scene images have different noise levels, which is used by TN for indoor/outdoor classification. The overall structure of the TN expert is shown in Fig. 3.20. First, TN computes a 9-dimensional noise feature vector for the whole image. Then, the SVM tool is used for training and testing.

Thermal Noise. Thermal noise [17] arises in the image acquisition process due to poor illumination, high temperature, etc. Typically, indoor scenes have weaker lighting sources and a lower TN value while outdoor scenes have stronger natural light and a higher TN value. For this reason, the TN value can be used to differentiate indoor/outdoor scenes.

First, we adopt a bilateral filter approach [38] to denoise each channel of the RGB, HSV and YUV color representations of an input image. The bilateral filter maintains the image edge structure well and provides robust and accurate de-noising results. As shown in Fig. 3.21, the absolute differences between the original and the denoised image channels are computed to yield 9 noise maps for one input image. Finally, the standard deviations of all noise maps are concatenated to form a 9-D feature vector. In our experiments, TN does not work as well as other experts, yet it does offer a supplemental solution.

HDH

The acronym, HDH, stands for the hue-dark histogram, where "dark" denotes the dark channel value of a pixel. The diagram of the HDH classification system is shown in Fig. 3.22. First, the image is divided into 4×4 blocks. Local HDH features are calculated for each block and, then, combined for SVM training and prediction. One of the main differences between HDH and traditional approaches (SP, SSL and PS) lies in the feature extraction step. We do not claim that the HDH expert works better than other experts on the average. However, HDH can provide more accurate performance on some images that cannot be well processed by others.

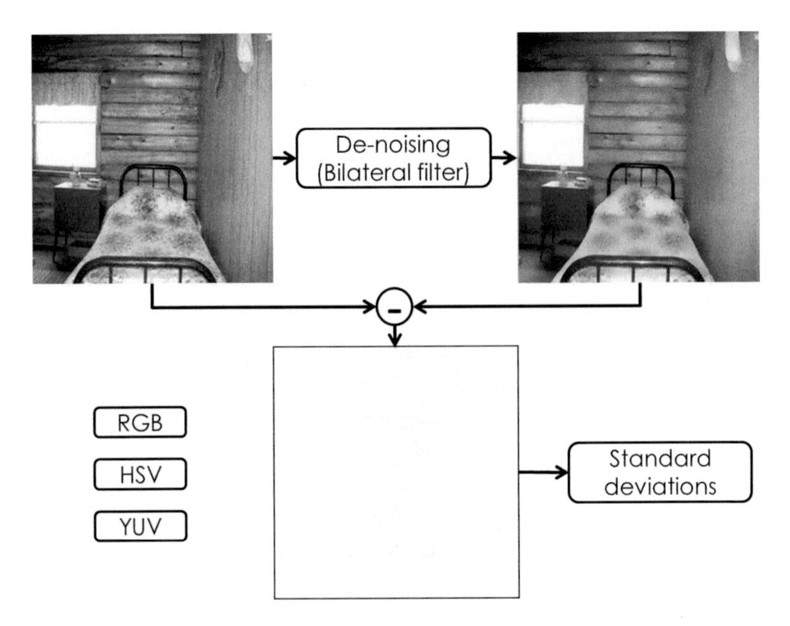

Fig. 3.21 Noise level computation for an input image

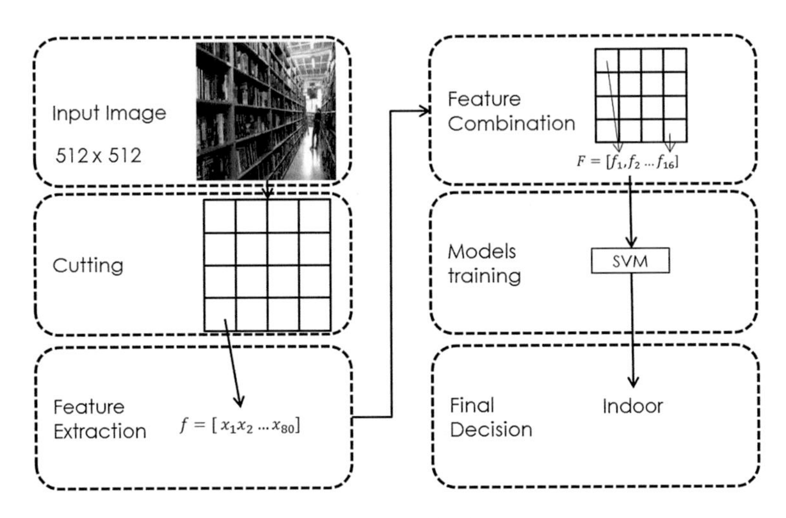

Fig. 3.22 An overview of the HDH indoor/outdoor classification system

Fig. 3.23 Comparison of *dark* channel values in *indoor* and *outdoor* scenes. **a** Original indoor library image. **b** Dark channel indoor library image. **c** Original outdoor library image. **d** Dark channel outdoor library image

The dark channel [14] of a pixel is defined to be the lowest one of its R, G, B three channel values and denoted by D. We observe that the dark channel values are different for indoor and outdoor scenes as shown in Fig. 3.23 due to different lighting sources and energy strength. This property cannot be well captured by the luminance value (of the Y, U, V three channels) of a pixel but the dark channel value.

Furthermore, we observe a different relationship between different colors and their dark channel values in indoor and outdoor images. To describe it, we adopt a 16-bin quantized Hue channel value and examine the local dark channel distribution with a 5-bin histogram for each Hue value. Thus, we obtain an 80-bin histogram feature for each block. Finally, we concatenate all block's feature vectors to a long one as the HDH descriptor of the whole image.

To conclude, HDH aims at modeling the distribution of pixel's darkest possible value for an image. The three introduced new experts, HSH, TN and HDH, serve can play a complementary role to the six traditional experts for indoor/outdoor classification.

3.3 Data Grouping Using Experts' Decisions

Data grouping is a widely adopted technique in statistics and machine learning. As millions of visual data, such as images and video clips, are created with the emergence of smart phones and mobile devices every day, an efficient data grouping scheme is important to visual data analytics. Besides, understanding the capabilities of machine-trained experts from the big visual data distribution aspect is crucial to an integrated smart system. In this section, we propose a novel method to group data according to the capabilities of different experts in the contexts of indoor/outdoor classification and vanishing point detection.

We select 18 images from the SUN database and show them in Fig. 3.24. They will be used as examples to facilitate our discussion below. Without loss of generality, we use expert KPK [19] as an illustrative example for indoor/outdoor scene classification. For the jth image sample, denoted by I_j, KPK can generate a soft decision score, d_j^{kpk}, for it using its sample-to-boundary distance normalized to the range of [0, 1], where 0 and 1 indicate the indoor and outdoor scenes with complete confidence, respectively. When there is only one expert, we need toquantize the soft decision score into a binary decision. That is, we divide interval [0, 1] into two subintervals $S_1 = [0, T)$ and $S_2 = [T, 1]$, where $0 < T < 1$ is a proper threshold value (typically, $T = 0.5$). If $d_j^{kpk} \in S_1$, I_j is classified to an indoor image. Otherwise, $d_j^{kpk} \in S_2$ and I_j is classified to an outdoor image.

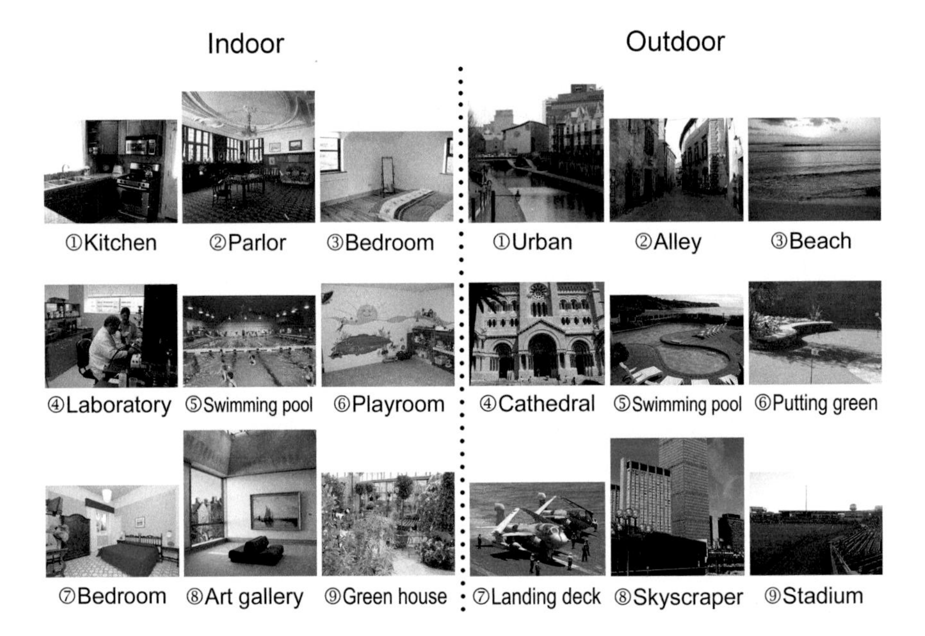

Fig. 3.24 Exemplary indoor and outdoor images in the SUN database

When the soft decision score, d_j^{kpk}, is closer to threshold T, expert KPK is less confident about its decision. To take this into account, we may partition the entire decision interval into 3 subintervals $S_1 = [0, T_1)$, $S_2 = [T_1, T_2)$, and $S_3 = [T_2, 1]$, where $0 < T_1 < T_2 < 1$ are two thresholds. Parameters T_1 and T_2 are set to 0.35 and 0.65 in our implementation. Subintervals S_1 and S_3 are called the confident regions while subinterval S_2 is called the uncertain region.

To gain more insights, we show the distribution of soft KPK decision scores collected from 5000 images randomly sampled from the SUN database in Fig. 3.25, where red circles and green crosses denote indoor and outdoor images, respectively. To avoid the overlap of cluttered samples along the x-axis, we generate a vertical random shift between −0.1 to 0.1 for each sample. This is purely for the visualization purpose and has no practical significance. We also provide the image population distribution and the KPK classification accuracy in each region in Table 3.9.

As shown in Fig. 3.25, we see a large number of red circles in S_1 and a large number of green crosses in S_3, which can be correctly classified by KPK. Correspondingly, higher correct classification rates in regions 1 and 3 are given in Table 3.9. There are significantly fewer red circles in S_3 and fewer green crosses in S_1, which are misclassified by KPK. There are some red circles and green crosses in S_2, which are difficult to set apart using soft KPK decision scores. This results in a lower classification rate in Table 3.9.

Based on the above discussion, the criteria of a good expert can be concluded as follows.

1. It has a larger ratio of correct versus incorrect decision samples in S_1 and S_3.
2. It has a smaller percentage of samples in S_2.

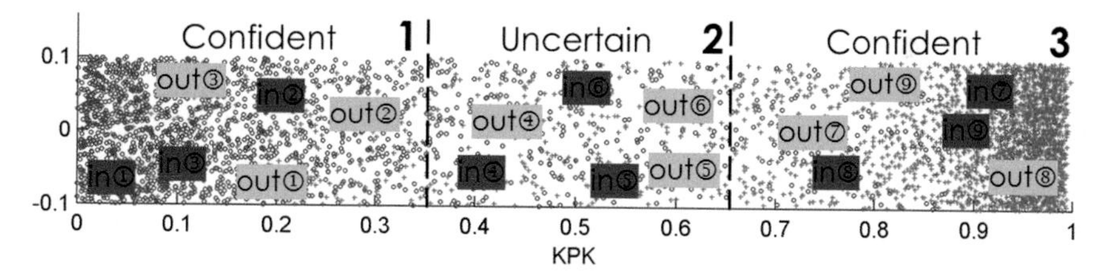

Fig. 3.25 The distribution of soft KPK decision scores d^{kpk} from 5,000 random samples

Table 3.9 The distribution of 5,000 randomly sampled images based on soft KPK scores and the corresponding correct indoor/outdoor classification rate in each region

Regions	Distribution (%)	KPK perforemance (%)
1	40.17	85.17
2	11.19	57.36
3	48.64	92.33
ALL	100.00	85.54

We will discuss ways to achieve these two goals by inviting the second expert to join the decision-making process in Sect. 3.4.1.

For the 18 images in Fig. 3.24, there are three representative indoor and outdoor images in each sub-interval in Fig. 3.25. Visual inspection of these sample images can help us understand the strength and weakness of KPK. **Images in the Uncertain Region**. For images in S_2, KPK cannot make a confident decision. Indoor images ④–⑥ and outdoor images ④–⑥ lie in this region. The two swimming pool images, images ⑤ in both indoor and outdoor categories, have similar elements such as blue water and dark tops. In addition, indoor and outdoor images ⑥ also share similar color patterns.

For images in S_1 and S_3, KPK has confident soft scores. They can be further divided into two cases. **Correctly Classified Images**. Indoor images ①–③ and outdoor images ⑦–⑨ are correctly classified. Recall that KPK partitions an image into 5 blocks (namely, one horizon block in the top and four parallel vertical blocks in the lower portion). Indoor images ①–③ have small d^{kpk} values since they all have red and wooden objects at the bottom part of the images and shell-white ceilings or walls, which are easy to classify with KPK's block-based color and edge descriptors. Similarly, the top horizontal block carries the valuable sky information for outdoor images ⑦–⑨.

Misclassified Images. Outdoor images ①–③ and indoor images ⑦–⑨ are misclassified although their scores fall in the confident regions. They are called outliers. Outdoor images ①–③ all have dark colors and clear edge structures over the entire image, which misleads KPK. The blue top part of indoor image ⑦ is also misleading. Indoor image ⑧ is difficult since its wall contains the outdoor view and painting. Indoor image ⑨ can be even challenging to human being since one may make a different decision depending on the existence of the ceiling and the wall.

For outlying images, low-level features mislead KPK to draw a confident yet wrong conclusion. Human can make a correct decision by understanding the semantic meaning of these scenes such as the river, the street and the ocean in outdoor images ①–③, respectively. Furthermore, indoor images ③ and ⑦ have the same semantic theme (bedroom) but different low-level features (color and texture patterns).

It is well known that there exists a gap between low-level features and high-level semantics knowledges of an image. This explains the fundamental limits of experts that rely purely on features in decision-making. Despite the semantic gap, a well-designed feature-based classifier can offer a reasonable classification performance due to the strong correlation between good low-level features and high-level semantics in a great majority of images.

3.4 Diversity Gain of Experts Via Decisions Stacking

As the size of image data becomes larger and their contents become more diversified, it is challenging to design a single expert that can handle all image types effectively. It is a natural idea to get the opinions of multiple experts and combine their opinions to form one final decision. In this section, we first consider the simplest two-expert case in Sect. 3.4.1. Then, we show the diversity gain provided by stacking the opinion of multiple experts in Sect. 3.4.2.

3.4.1 Diversity Gain of Two Experts

To get the opinions of two experts and improve the classification accuracy, we simply stack two experts' decisions. Intuitively, such a decision stacking system [8, 9, 45] may not work well under the following two scenarios: (1) if the opinions of two experts are too similar to each other; or (2) if one expert is significantly better than the other. In both scenarios, we do not benefit much by inviting the second expert in the decision process. Scenario (2) is self-evident. We will focus on scenario (1) by investigating the diversity gain of a two-expert system.

Without loss of generality, we choose KPK [19] and PY [29] as two experts for indoor/outdoor classification. The soft decision scores of expert PY, denoted by d^{py}, for the same 5,000 samples (as plotted in Fig. 3.25) are plotted along the vertical axis in Fig. 3.26. The jth sample image represented by a red circle (indoor) or a green cross (outdoor) has two soft scores, denoted by (d_j^{kpk}, d_j^{py}), which defines the KPK-PY soft decision map. With different combinations of soft decisions from two experts, we can divide the 2-D decision space into 9 regions as shown in Fig. 3.26. KPK and PY have consistent opinions in their soft decisions in regions 1, 5 and 9, complementary decisions in regions 2, 4, 6 and 8, and contradictory opinions in regions 3 and 7.

Table 3.10 shows the distribution of the 5000 images and the correct classification rate of expert KPK and expert PY, respectively, in each region. KPK and PY have similar classification rates in the consistent regions (i.e., regions 1, 5 and 9). One of them outperforms the other in the complementary regions (i.e. regions 2, 4, 6 and 8). The sum of their decisions is equal to unity in the contradictory regions (i.e. regions 3 and 7) since one of them should be correct in the binary decision case.

By comparing Figs. 3.25 and 3.26, we see that PY can help KPK in resolving some decision ambiguities in regions 4 and 6. Similarly, KPK can help PY in resolving some decision ambiguity in regions 2 and 8. KPK and PY offer complementary strength since they examine different low-level features in evaluating an input image. KPK focuses on the local color and edge distributions while PY focuses on global scene structures. In Fig. 3.24, indoor/outdoor images ②, ④, ⑥ and ⑧ are exemplary images in regions 2, 4, 6 and 8, respectively.

Consider indoor/outdoor images ④ and ⑥, for which KPK does not have a confident score. Recall that PY [29] does not partition an image into multiple sub-images but computes the GIST [28] features from the original image and its edge map separately and cascades the two responses into a feature vector. As a result, PY can make a more confident decision. PY's decisions on indoor image ④ (complicated scene structure for the whole image) and outdoor image ⑥ (textures of grass and leaves) are correct, yet PY's decisions on indoor image ⑥ (similar to outdoor

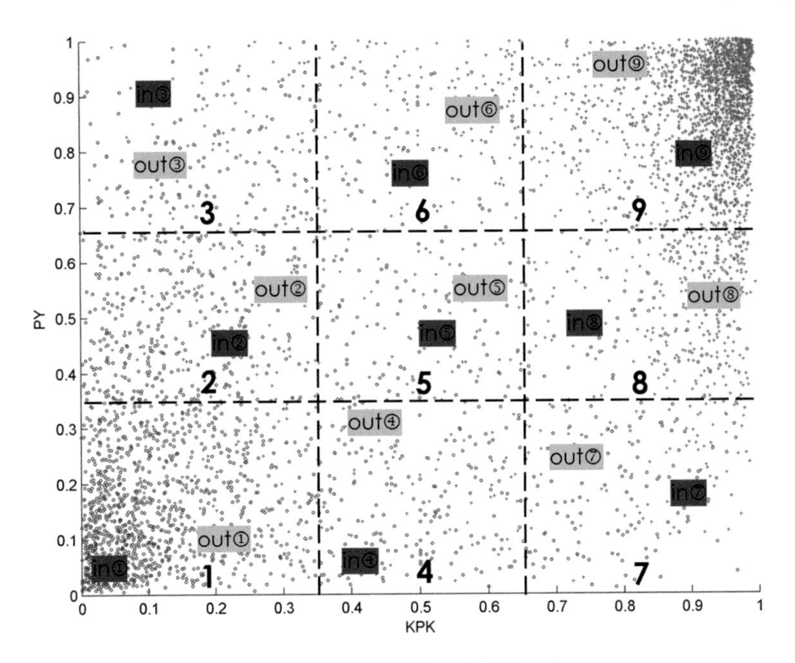

Fig. 3.26 Diversity gain in stacking soft decisions of KPK and PY

Table 3.10 The distribution of 5,000 randomly sampled images based on soft score pairs of KPK and PY and the corresponding correct indoor/outdoor classification rates of KPK and PY in each region

Regions	Distribution (%)	KPK performance (%)	PY performance (%)
1	28.72	90.75	90.75
2	7.22	78.86	60.03
3	4.23	58.08	41.92
4	4.18	57.91	68.53
5	2.81	57.56	55.40
6	4.20	56.68	76.64
7	4.17	58.34	41.66
8	7.47	86.66	59.79
9	37.00	97.31	97.31
ALL	100.00	85.54	82.02

image ⑥) and outdoor image ④ (consisting of many straight vertical lines similar to the view observed inside a church building) are not accurate. Since there are more indoor images than outdoor images in region 4 and more outdoor images than indoor images in region 6, PY does contribute to the correct classification rate in regions 4 and 6. The same discussion applies to regions 2 and 8, where KPK helps PY in resolving the ambiguity in a positive way.

Region 5 remains to be ambiguous in the two-expert system. If the two experts share very similar opinions, most samples will fall in regions 1, 5 and 9 and, as a result, the two-expert system does not offer a clear advantage. However, if the two experts have good but diversified opinions, we will observe more samples in the four complementary regions and the overall classification performance can be improved.

Finally, PY and KPK have conflicting opinions in regions 3 and 7. To resolve the conflict, we can invite other experts in the decision making process as detailed in Sect. 3.4.2.

In the discussion above, we choose KPK as the first candidate since it provides the highest classification accuracy over the SUN database. PY is selected as the second candidate since it provide an excellent complementary role to KPK in our experiment. In addition, by stacking soft decisions of each of possible pair of the 9 indoor/outdoor classification experts and training a meta-level classifier, we are able to show and compare diversity gains achieved by different pairs of experts in Table 3.11, where the 5-fold cross validation is conducted on the same 5,000 random sample images selected earlier, and the averaged classification accurate rates are shown.

The numbers along the diagonal line are individual classification rates of single-expert systems while non-diagonal numbers are the classification rates of two-expert systems. Among the individual experts, KPK provides the best performance, which is 85.84 % (in red). KPK has its "best" partner PY, which improves KPK's performance from 85.84 to 87.22 %. However, the pair of KPK and PY does not provide the best two-expert system on the 5,000 image samples. The pair of PY and HDH yields the best two-expert system that reaches a classification rate of 87.68 %. Although HDH does not perform as well as KPK. This is because that HDH is more complementary to PY than KPK on the 5,000 images.

Table 3.11 The averaged classification accurate rates for various two-expert systems, where the 5-fold cross validation is conducted on the same 5,000 random sample images selected earlier

	SP [37]	VFJ [42]	SSL [34]	PY [29]	KPK [19]	XHE [47]	TN	HSH	HDH
SP	83.64	83.76	85.20	87.58	86.32	84.20	85.20	84.10	84.94
VFJZ	–	80.58	85.20	85.94	85.78	80.92	81.88	80.90	82.52
SSL	–	–	85.10	87.62	86.78	85.38	86.20	85.54	86.00
PY	–	–	–	82.18	87.22	86.62	82.48	86.18	87.68
KPK	–	–	–	–	85.84	85.86	86.28	85.78	85.84
XHEOT	–	–	–	–	–	80.70	82.38	82.18	84.18
TN	–	–	–	–	–	–	63.58	80.58	83.68
HSH	–	–	–	–	–	–	–	78.86	82.30
HDH	–	–	–	–	–	–	–	–	82.44

3.4.2 Construction of Multi-Expert Systems

Generally speaking, we can build a multi-expert system by adding experts one by one with one of the following two strategies.

Best expert first. With this strategy, we choose the expert in the remaining ones that offers the best individual performance. To begin with, KPK is the best one among the 9 experts in Table 3.11, it is used in the single-expert system. Then, SSL is selected as the second one, SP as the third one, and so on. We conduct the experiment on the 5,000 images using the 5-fold cross validation and plot the correct classification rate curve as a function of the number of experts using the "best expert first" strategy in Fig. 3.27. As shown in the figure, when all experts participate in the decision fusion process, the final performance accuracy reaches 89.14 %.

Most complimentary experts first. To begin with, we select the first two most complimentary experts, PY and HDH, based on the results in Table 3.11. Then, to select the next one, we choose the one in the remaining candidates that has the best complimentary strength to the existing system. That is, we perform the 5-fold cross validation on the selected 5,000 images by adding the third expert one by one from the remaining list. It turns out that KPK provides the best classification accuracy among the remaining 7 experts when its decision is fused with those of PY and HDH. The same process is repeated until all experts are selected. The order of selected experts and the performance of the multi-expert system are shown in Fig. 3.28. The performance curve in Fig. 3.28 rises up quickly when adding a few capable experts in the beginning, but it drops slightly after the seven-expert system (consisting of PY, HDH, KPK, SSL, HSH, XHE and VFJ). The seven-expert system can achieve

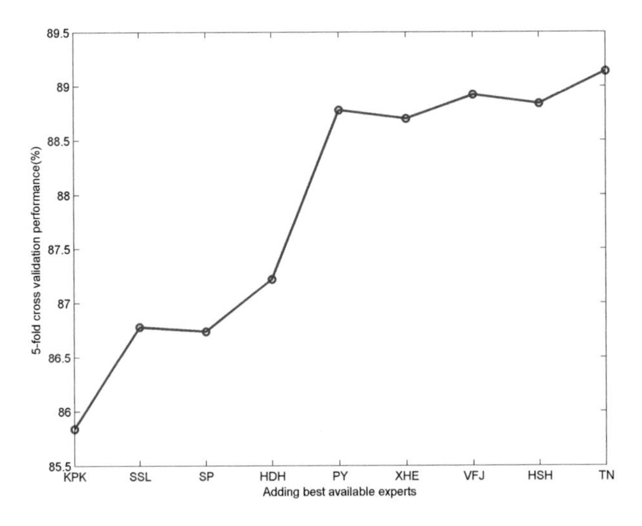

Fig. 3.27 The classification rate performance as a function of the expert number using the best-expert-first strategy, where the x-axis shows the new expert to be added at each round

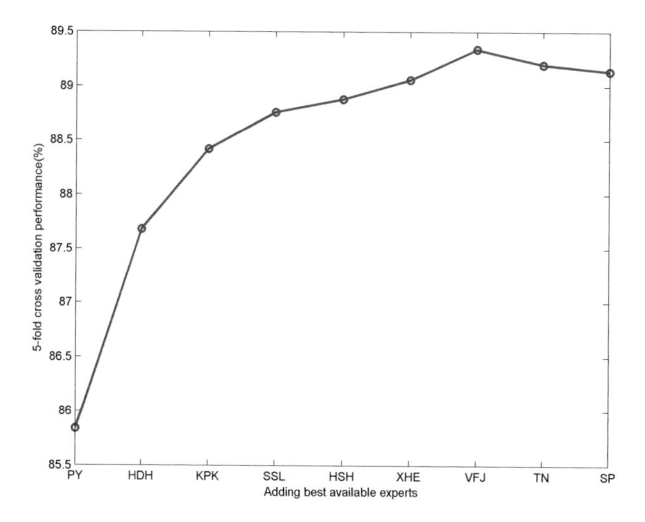

Fig. 3.28 The classification rate performance as a function of the expert number using the most-complementary-first strategy, where the x-axis shows the new expert to be added at each round

a correct classification rate of 89.34 %, which is better than the performance of the nine-expert system. This is attributed to the overfitting problem in machine learning.

When the number of experts becomes larger, their weights (and, thus, performance) are more sensitive to the size of the training data. It is also worthwhile to point out that our discussion in above applies to the selected 5,000 images, and the performance behavior may change with respect to a different dataset.

3.5 Expert Decision Fusion Systems

The block diagram of an Expert Decision Fusion (EDF) system is shown in Fig. 3.29, which consists of two stages: (1) grouping and (2) decision fusion via stacking.

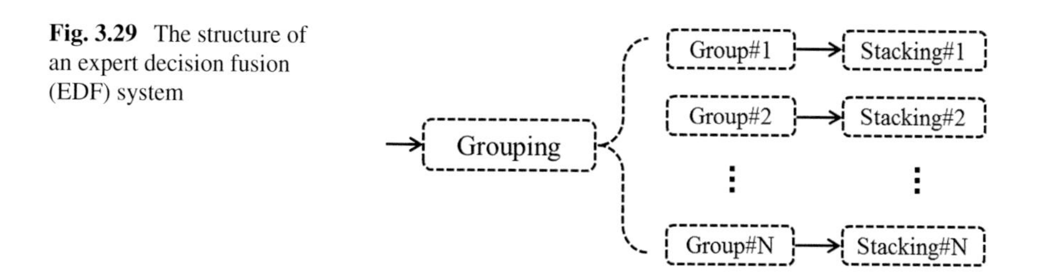

Fig. 3.29 The structure of an expert decision fusion (EDF) system

At the first stage, we perform data grouping based on the soft score of a single expert or soft scores of multiple experts. Given experts' soft decision map, we partition the data space into regions. Generally speaking, the data grouping scheme provides a powerful pre-processing step in machine learning. The main purpose is to increase the correlation between training and test data samples. A good grouping strategy can contribute to the overall performance of the learning-based system significantly. After grouping, the diversity of data samples in each region is reduced.

At the second stage, we fuse soft decisions of all nine experts in each group by forming a meta-level classification model. One simple fusion method is voting. That is, we binarize the soft decision of each expert and use the simple majority voting rule to fuse expert's decisions. A more advanced fusion method is stacking, where we build a meta-level classification model that takes soft scores of all experts as the input features and make a final binary system decision. Since the training data in each group are different, different meta-level data models are built for different groups. The meta-level classifier is trained by linear SVM using samples with known binary outputs and, then, the trained model is used to predict samples with unknown binary outputs in the test. Usually, the stacking scheme provides a better fusion result.

With a good grouping stage, the EDF system can be very powerful when dealing with large-scale visual data. We discussed one-expert grouping and two-expert grouping schemes in the last section. The whole EDF system with one-expert grouping and two-expert grouping schemes are shown in Figs. 3.30 and 3.31, respectively. As shown in these figures, we can conduct decision stacking based on the soft scores of all nine experts in each region.

It is possible to extend the group idea to include the soft scores of more experts. For example, we show a three-expert grouping system in Fig. 3.32 based on the joint soft scores of KPK, PY and XHE, where the data space is partitioned into 27 subspaces. Note that XHE is selected because it is the best team player for KPK and PY among the remaining experts.

Both grouping and stacking provide powerful tools to handle the data diversity problem. Through grouping, we have more and smaller homogeneous datasets rather than one large highly heterogeneous dataset. Through stacking, we can improve the robustness of the final decision in each region by leveraging the complementary strength of multiple experts.

3.6 Performance Evaluation

3.6.1 Performance of Individual Expert

Indoor/outdoor classification on SUN. We conduct 5-fold cross validation for each expert on the whole SUN dataset. The full set of 108,754 images is randomly partitioned into five parts of similar sizes, where each part has about 21,750 images. For each test, four of the five parts are used as training data and the remaining one part

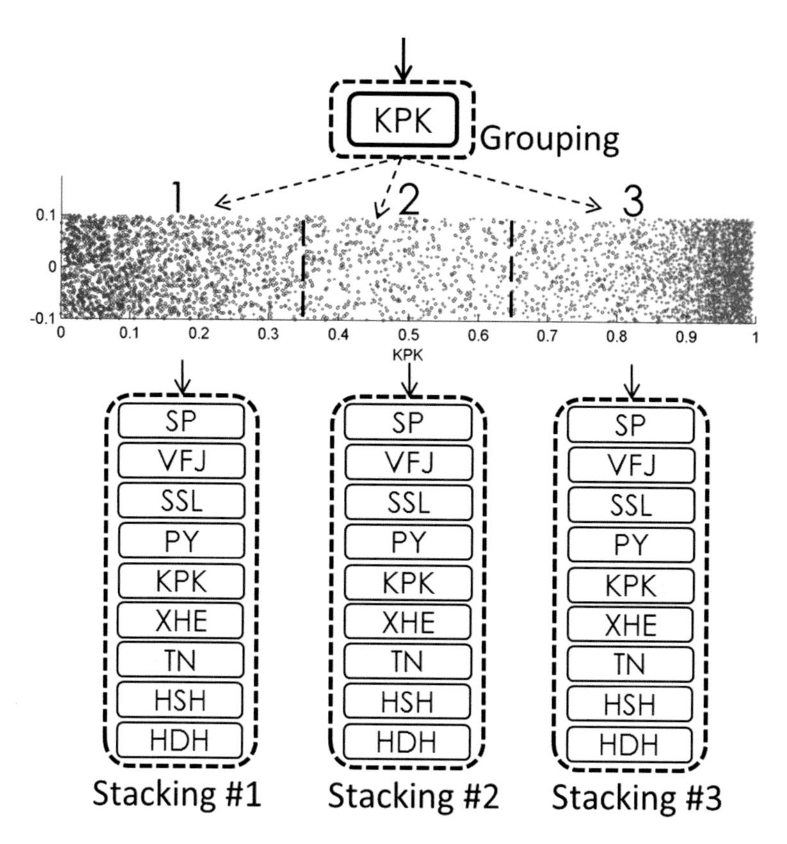

Fig. 3.30 The indoor/outdoor scene classification EDF system with one-expert grouping (based on the KPK soft score)

is used as the test data. The correct classification results of the 5-fold cross validation are averaged and used as the performance measure of each expert. We list the performance of each of the nine experts in Table 3.12.

As shown in Table 3.12, the nine experts have a performance ranging from 64.21 to 85.53 %. All experts proposed before (from SP to XHE) perform worse than the claim in their original papers by 5–11 % since the datasets used in their original work were smaller and without the same diversity as that of the SUN dataset. Among all indoor/outdoor classification experts, KPK gives the best result in our experiment so that we use its soft score as the basis of grouping in the one-expert grouping scheme.

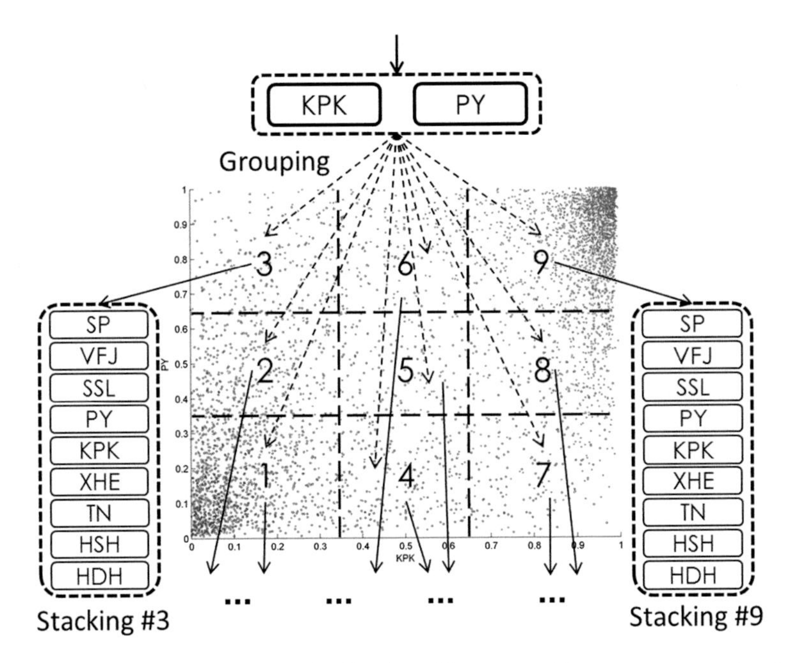

Fig. 3.31 The indoor/outdoor scene classification EDF system with two-expert grouping (based on the joint soft scores of KPK and PY)

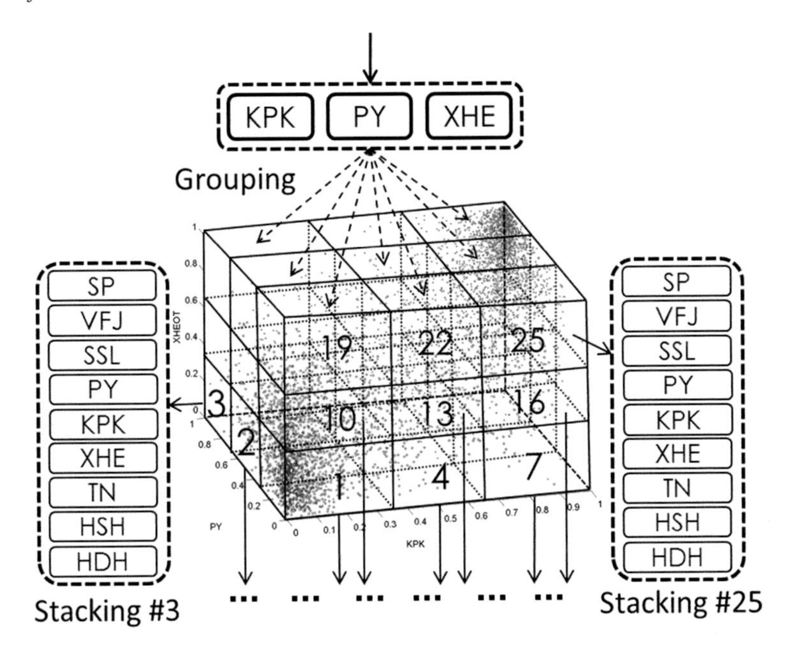

Fig. 3.32 The indoor/outdoor scene classification EDF system structure with three-expert grouping (based on the joint soft scores of KPK, PY and XHE)

Table 3.12 Performance of the nine indoor/outdoor classification experts on the entire SUN dataset

SP	VFJ	SSL	PY	KPK	XHE	TN	HSH	HDH
82.68 %	79.23 %	84.66 %	82.21 %	85.53%	78.95 %	64.21 %	77.97 %	82.30 %

3.6.2 Subspace Classification Performance

Three EDF systems were presented in Sect. 3.5, where their difference lies in different subspaces after grouping. We use the whole SUN dataset to evaluate the performance of the three EDF systems. We focus on indoor/outdoor EDF systems in this section.

First, we show the correct classification rate of the EDF system using the one-expert grouping scheme and decision stacking of nine experts in Regions 1-3 in Table 3.13, where the 5-fold cross validation is conducted in each region and the averaged performance is reported. For the last row, we perform the weighted sum of correct rates in nine regions based on their sample distribution. For EDF_V, we binarize the soft decision of each expert and use the majority voting scheme to fuse experts' decisions. For EDF_S, we adopt the stacking scheme to fuse experts' decisions. That is, we treat experts' decisions as features and build a meta-level SVM classifier on top of them. We see that the performance of EDF_S is no worse than EDF_V in all regions for the EDF system. Thus, it is chosen to be the final EDF solution.

By comparing results in Tables 3.12 and 3.13, we see clearly that the performance of each expert has been improved a lot (ranging from 4 to 22 %) due to data grouping by KPK. After data grouping, the performance gap among different experts narrows down significantly. Their correct classification rates now are in the range of 85.22–87.53 %. The EDF system can achieve a correct classification rate of 88.60 % by stacking all experts in each region. With the combination of grouping and stacking, the EDF system can outperform individual experts (without grouping) by a margin of 3–24 %.

We follow the same procedure of the one-expert EDF system and conduct experiments for the two-expert and the three-expert grouping EDF systems. The results of these two systems are shown in Tables 3.14 and 3.15, respectively.

We observe even more improvement in Table 3.14 by dividing data samples into nine regions based on the joint soft scores of KPK and PY. The partitioning of the

Table 3.13 Performance of the one-expert grouping EDF system on the SUN dataset for indoor/outdoor classification

	SP	VFJ	SSL	PY	KPK	XHE	TN	HSH	HDH	EDF_V	EDF_S	EDF
1	86.93	86.92	86.93	86.92	86.92	86.92	86.96	86.93	86.94	86.92	87.43	87.43
2	64.46	54.25	67.35	71.01	61.17	61.16	64.92	63.50	63.51	69.34	75.82	75.82
3	90.74	90.90	90.86	91.69	90.89	90.90	90.89	90.91	90.89	90.90	92.34	92.34
All	86.30	85.22	86.69	87.53	86.00	86.00	86.44	86.28	86.27	86.94	88.60	88.60

Table 3.14 Performance of the two-expert grouping EDF system on the SUN dataset for indoor/outdoor classification

	SP	VFJ	SSL	PY	KPK	XHE	TN	HSH	HDH	EDF_V	EDF_S	EDF
1	92.22	92.22	92.22	92.22	92.22	92.22	92.22	92.22	92.22	92.22	92.43	92.43
2	76.91	75.99	78.24	80.07	75.96	75.96	75.92	75.94	77.34	75.96	86.32	86.32
3	71.63	61.18	73.18	70.07	66.17	73.18	61.47	66.75	70.49	77.48	81.36	81.36
4	83.46	83.22	83.67	83.22	83.22	83.22	83.22	83.32	83.41	83.22	85.58	85.58
5	68.96	56.87	70.50	65.21	60.23	70.16	61.57	66.14	67.79	68.23	79.84	79.84
6	81.17	81.07	81.72	80.90	81.09	80.99	80.87	81.06	81.70	80.90	86.32	86.32
7	70.04	63.06	71.72	68.68	66.14	71.06	63.06	64.62	68.80	65.60	80.12	80.12
8	72.51	70.56	72.58	75.35	70.37	72.44	70.35	70.58	72.05	70.51	81.54	81.54
9	96.70	96.74	96.74	96.75	96.74	96.74	96.74	96.74	96.74	96.74	96.74	96.74
All	88.64	87.44	88.96	88.64	87.87	88.73	87.56	88.00	88.56	88.52	91.15	91.15

entire data space into nine regions narrows down data diversity furthermore and contribute to classification improvement for all experts. By stacking all experts' decisions in each region, the EDF system can outperform individual experts by a margin of 6–26 % with a final correct classification rate of 91.15 % on the whole SUN dataset.

However, by comparing results in Tables 3.15 and 3.14, we see that the three-expert grouping EDF system does not outperform the two-expert grouping EDF system. The main reason is that there is no sufficient data in each partitioned region after the three-expert grouping system. In the experiment, for every fold of the cross validation, we have aroung 87,000 images in the training dataset. Most of them are in Regions 1 and 27 after grouping, and experts' decisions in these two regions are very consistent. There is little improvement in diversity reduction by moving from the two-expert grouping to the three-expert grouping. On the other hand, the number of training images becomes much less from regions 2 to 26. The overfitting problem starts to appear in these regions. It is concluded from the above discussion that the two-expert grouping EDF system is most suitable for the SUN dataset.

3.6.3 Scalability

We plot the performance of each individual indoor/outdoor expert (without data grouping) and the indoor/outdoor EDF system (two-expert grouping EDF) as a function of the size of the dataset in Fig. 3.33, where we select subsets of increasing sizes randomly selected from the SUN dataset and list the data size in the x-axis. Furthermore, the averaged correct classification rate using the 5-fold cross validation is shown in the y-axis. The vertical segment on each marker indicates the standard deviation of a particular test. We see that the performance of some individual experts

Table 3.15 Performance of the three-expert grouping EDF system on the SUN dataset for indoor/outdoor classification

	SP	VFJ	SSL	PY	KPK	XHE	TN	HSH	HDH	EDF_V	EDF_S	EDF
1	94.31	94.31	94.30	94.31	94.31	94.31	94.30	94.31	94.31	94.31	94.11	94.31
2	81.56	81.56	81.56	82.06	81.56	81.56	81.56	81.63	81.46	81.56	80.78	81.56
3	71.16	71.86	73.14	73.44	73.23	71.97	72.15	73.65	72.38	72.18	75.96	75.96
4	87.82	87.82	87.72	87.82	87.82	87.82	87.82	87.78	87.76	87.82	86.45	87.82
5	63.73	63.82	65.03	65.64	64.30	64.17	64.01	65.37	65.95	63.91	69.84	69.84
6	62.97	63.27	67.06	71.17	68.09	62.55	63.29	67.45	66.64	65.21	72.05	72.05
7	73.93	73.90	73.97	75.21	73.87	73.90	74.11	73.85	74.28	73.90	74.56	74.56
8	57.59	57.96	60.35	66.37	66.59	59.59	60.71	58.88	60.95	60.47	67.54	67.54
9	85.43	85.96	85.43	87.07	86.83	85.96	85.96	86.06	85.63	85.96	85.16	85.96
10	90.80	74.40	75.19	79.89	78.55	90.82	74.40	74.41	74.46	74.54	90.86	90.86
11	75.52	75.51	76.05	77.67	75.57	75.57	75.36	75.56	75.53	75.57	77.35	77.35
12	67.30	64.90	68.41	69.22	66.29	65.10	65.30	67.82	68.17	65.85	75.47	75.47
13	78.43	77.93	78.74	78.75	77.81	77.93	78.43	77.92	78.21	77.93	79.72	79.72
14	63.19	59.38	64.25	61.82	57.99	59.21	63.35	61.99	63.64	61.44	69.08	69.08
15	68.74	68.76	70.22	68.75	70.03	68.58	68.53	71.00	69.59	68.76	73.93	73.93
16	63.51	58.33	65.77	63.01	60.16	58.81	66.26	59.61	61.82	61.54	70.21	70.21
17	69.70	70.48	69.61	71.43	71.00	70.48	70.48	70.34	70.17	70.48	72.18	72.18
18	92.06	92.11	92.09	92.09	92.09	92.09	92.09	92.08	92.09	92.09	91.29	92.09
19	82.38	82.51	82.41	83.60	82.51	82.51	82.62	82.45	82.37	82.51	82.81	82.81
20	69.07	66.47	71.25	71.52	66.11	66.28	66.28	66.04	71.28	66.82	75.95	75.95
21	68.33	63.77	70.08	69.12	67.97	67.93	64.73	67.92	68.60	69.05	76.66	76.66
22	65.28	61.59	68.27	65.18	62.58	62.61	64.48	65.33	64.33	65.86	67.47	67.47
23	62.44	58.02	65.40	59.39	58.47	58.44	58.03	62.48	60.54	58.60	67.05	67.05
24	88.69	88.73	88.75	88.73	88.75	88.73	88.73	88.67	88.69	88.73	86.87	88.73
25	70.50	69.11	70.45	71.18	70.86	69.48	69.11	68.25	70.74	69.77	73.82	73.82
26	87.36	87.59	87.59	87.59	87.59	87.59	87.59	87.68	87.60	87.59	83.82	87.59
27	98.35	98.35	98.35	98.35	98.35	98.35	98.35	98.34	98.35	98.35	98.34	98.35
All	88.19	86.65	87.31	87.79	87.31	88.02	86.90	86.97	87.10	86.95	89.01	89.32

stays flat while that of others drops as the data size becomes larger. In contrast, the performance of the EDF system keeps improving as the data size increases.

When the data size becomes larger, there are two competing factors that influence the performance of a classification system in two opposite directions. On one hand, the data become more diversified and the performance of an expert may go down if its model cannot handle a diversified data set. On the other hand, the number of similar data (i.e., belonging to the same data type) becomes more. Data abundance helps improve the performance of learning-based classifiers. Figure 3.33 implies that the data diversity problem is under control by the robust EDF system so that the EDF

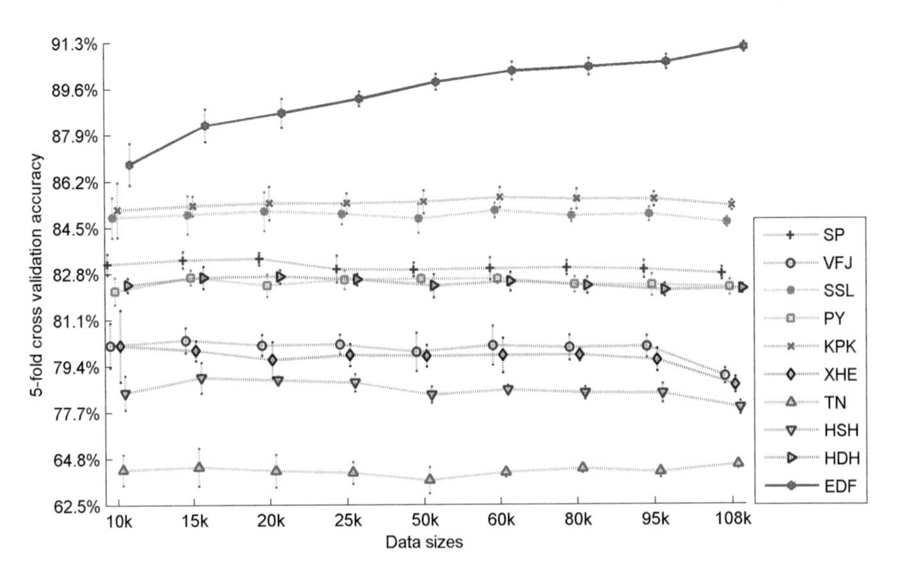

Fig. 3.33 Comparison of indoor/outdoor classification performance of nine indoor/outdoor experts and indoor/outdoor EDF as a function of the size of dataset

system benefits more from data abundance. We expect the EDF system performance to be level-off at a certain data size (although we have not yet observed such a phenomenon in Fig. 3.33) since that the EDF system does not address the semantic gap issue.

3.6.4 Discussion

We can see from Table 3.16 that the EDF system performs the best and achieves a correct classification rate of 91.15 % for the indoor/outdoor classification problem. With the combination of grouping and stacking, the EDF system can outperform traditional experts (without grouping) by a margin of 6–26 %.

To conclude, the EDF system outperforms all individual experts in dealing with the indoor/outdoor classification problem. It offers a valuable solution to the big visual data classification problem by integrating the capabilities of individual classification experts in a smart way.

Table 3.16 Performance of nine individual indoor/outdoor experts and the corresponding EDF system

SP	VFJ	SSL	PY	KPK	XHE	TN	HSH	HDH	EDF
82.68 %	79.23 %	84.66 %	82.21 %	85.53 %	78.95 %	64.21 %	77.97 %	82.30 %	91.15 %

3.7 Summary

An Expert Decision Fusion (EDF) system was developed to address the large-scale indoor/outdoor image classification problem in this proposal. As compared with the traditional classifiers (or experts), the EDF system consists of two key ideas: (1) grouping of data samples based on the soft decisions of one or multiple experts into several sub-regions; and (2) stacking of soft decisions from all constituent experts to enhance the classification performance in each region. It was shown by experimental results that the proposed EDF system with 2 base experts outperforms all traditional classifiers in the classification accuracy by a huge margin of 6–26 % on the large-scale SUN image dataset. In addition, the classification performance of EDF improves as the size of the dataset grows, which can be explained by its capability of handling data diversities. With this capability in place, as the dataset grows to a very large size, data abundance becomes a more dominant factor than data diversity. Thus, the EDF system offers a robust and scalable solution.

As discussed in Sect. 3.5, there is some fundamental limits in the low-level feature-based classifiers since they do not take the image semantics into account. We expect to see a point where the performance of EDF becomes saturated, which will be the true upper performance bound of EDF. To achieve this goal, we need to look for some dataset even larger than SUN that has indoor/outdoor labels. To improve the performance of EDF furthermore beyond the saturation point, we need to look for semantic-based experts. This is clearly a very challenging problem since it involves object and scene recognition. Finally, a good indoor/outdoor image classifier is an important pre-processing step to scene analysis. It is desirable to leverage our current results to obtain better methods for scene classification and recognition.

References

1. Battiato, S., Curti, S., Cascia, M.L., Tortora, M., Scordato, E.: Depth map generation by image classification. Proc. SPIE **5302**, 95–104 (2004)
2. Bianco, S., Ciocca, G., Cusano, C., Schettini, R.: Improving color constancy using indoor-outdoor image classification. IEEE Trans. Image Process. **17**(12), 2381–2392 (2008)
3. Bosch, A., Zisserman, A., Muñoz, X.: Scene classification via plsa. In: Computer Vision-ECCV 2006. Springer, New York (2006), pp. 517–530
4. Boutell, M.R., Luo, J., Shen, X., Brown, C.M.: Learning multi-label scene classification. Pattern Recognit. **37**(9), 1757–1771 (2004)
5. Chang, C.C., Lin, C.J.: Libsvm: a library for support vector machines. ACM Trans. Intell. Syst. Technol. (TIST) **2**(3), 27 (2011)
6. Cortes, C., Vapnik, V.: Support-vector networks. Mach. Learn. **20**(3), 273–297 (1995)
7. Daubechies, I.: Ten Lectures on Wavelets, vol. 61. SIAM (1992)
8. Deng, L., Yu, D., Platt, J.: Scalable stacking and learning for building deep architectures. In: IEEE International Conference on Acoustics, Speech and Signal Processing (ICASSP), 2012, pp. 2133–2136. IEEE (2012)
9. Džeroski, S., Ženko, B.: Is combining classifiers with stacking better than selecting the best one? Mach. Learn. **54**(3), 255–273 (2004)

10. Fei-Fei, L., Perona, P.: A bayesian hierarchical model for learning natural scene categories. In: IEEE Computer Society Conference on Computer Vision and Pattern Recognition, 2005. CVPR 2005, vol. 2, pp. 524–531. IEEE (2005)
11. Fogel, I., Sagi, D.: Gabor filters as texture discriminator. Biol. Cybern. **61**(2), 103–113 (1989)
12. Gonzalez, R.C., Woods, R.E.: Digital Image Processing. Prentice Hall, New Jersey (2002), pp. 462–463
13. Gray, R.M., Olshen, R.A.: Vector quantization and density estimation. In: Proceedings on Compression and Complexity of Sequences 1997. IEEE (1997), pp. 172–193
14. He, K., Sun, J., Tang, X.: Single image haze removal using dark channel prior. IEEE Trans. Pattern Anal. Mach. Intell. **33**(12), 2341–2353 (2011)
15. Huang, J., Liu, Z., Wang, Y.: Joint scene classification and segmentation based on hidden markov model. IEEE Trans. Multim. **7**(3), 538–550 (2005)
16. Jégou, H., Douze, M., Schmid, C.: Improving bag-of-features for large scale image search. Int. J. Comput. Vis. **87**(3), 316–336 (2010)
17. Johnson, J.B.: Thermal agitation of electricity in conductors. Phys. Rev. **32**(1), 97 (1928)
18. Karpenko, A., Aarabi, P.: Tiny videos: Non-parametric content-based video retrieval and recognition. In: Tenth IEEE International Symposium on Multimedia, 2008. ISM 2008. IEEE (2008), pp. 619–624
19. Kim, W., Park, J., Kim, C.: A novel method for efficient indoor-outdoor image classification. J. Signal Process. Syst. **61**(3), 251–258 (2010)
20. Kohonen, T., Kangas, J., Laaksonen, J., Torkkola, K.: Lvq_pak: a software package for the correct application of learning vector quantization algorithms. In: International Joint Conference on Neural Networks, 1992. IJCNN, vol. 1. IEEE (1992), pp. 725–730
21. Lazebnik, S., Schmid, C., Ponce, J.: Beyond bags of features: spatial pyramid matching for recognizing natural scene categories. In: IEEE Computer Society Conference on Computer Vision and Pattern Recognition, 2006, vol. 2. IEEE (2006), pp. 2169–2178
22. Li, L.J., Su, H., Fei-Fei, L., Xing, E.P.: Object bank: A high-level image representation for scene classification and semantic feature sparsification. In: Advances in Neural Information Processing Systems, pp. 1378–1386 (2010)
23. Liu, C., Yuen, J., Torralba, A.: Nonparametric scene parsing: label transfer via dense scene alignment. In: IEEE Conference on Computer Vision and Pattern Recognition, 2009. CVPR 2009. IEEE (2009), pp. 1972–1979
24. Maini, R., Aggarwal, H.: Study and comparison of various image edge detection techniques. Int. J. Image Process. (IJIP) **3**(1), 1–11 (2009)
25. Mao, J., Jain, A.K.: Texture classification and segmentation using multiresolution simultaneous autoregressive models. Pattern Recognit. **25**(2), 173–188 (1992)
26. Murphy, K., Torralba, A., Freeman, W., et al.: Using the forest to see the trees: a graphical model relating features, objects and scenes. Adv. Neural Inf. Process. Syst. **16**, 1499–1506 (2003)
27. Ohta, Y.I., Kanade, T., Sakai, T.: Color information for region segmentation. Comput. Graph. Image Process. **13**(3), 222–241 (1980)
28. Oliva, A., Torralba, A.: Modeling the shape of the scene: a holistic representation of the spatial envelope. Int. J. Comput. Vis. **42**(3), 145–175 (2001)
29. Pavlopoulou, C., Yu, S.X.: Indoor-outdoor classification with human accuracies: Image or edge gist? In: IEEE Computer Society Conference on Computer Vision and Pattern Recognition Workshops (CVPRW), 2010. IEEE (2010), pp. 41–47
30. Picard, R.W., Kabir, T., Liu, F.: Real-time recognition with the entire brodatz texture database. In: IEEE Computer Society Conference on Computer Vision and Pattern Recognition, 1993. Proceedings CVPR'93, 1993. IEEE (1993), pp. 638–639
31. Quattoni, A., Torralba, A.: Recognizing indoor scenes. In: IEEE Conference on Computer Vision and Pattern Recognition (CVPR) (2009)
32. Rissanen, J.: Stochastic complexity. J. R. Stat. Soc. Ser. B (Methodological), 223–239 (1987)
33. Russell, B.C., Torralba, A., Murphy, K.P., Freeman, W.T.: Labelme: a database and web-based tool for image annotation. Int. J. Comput. Vis. **77**(1–3), 157–173 (2008)

34. Serrano, N., Savakis, A., Luo, A.: A computationally efficient approach to indoor/outdoor scene classification. In: 16th International Conference on Pattern Recognition, Proceedings, vol. 4, pp. 146–149 (2002)
35. Stella, X.Y., Zhang, H., Malik, J.: Inferring spatial layout from a single image via depth-ordered grouping. In: CVPR Workshop (2008)
36. Swain, M.J., Ballard, D.H.: Color indexing. Int. J. Comput. Vis. **7**(1), 11–32 (1991)
37. Szummer, M., Picard, R.W.: Indoor-outdoor image classification. In: IEEE International Workshop on Content-Based Access of Image and Video Database on, pp. 42–51. IEEE (1998)
38. Tomasi, C., Manduchi, R.: Bilateral filtering for gray and color images. In: Sixth International Conference on Computer Vision, 1998. IEEE (1998), pp. 839–846
39. Torralba, A., Fergus, R., Freeman, W.T.: 80 million tiny images: a large data set for nonparametric object and scene recognition. IEEE Trans. Pattern Anal. Mach. Intell. **30**(11), 1958–1970 (2008)
40. Vailaya, A., Figueiredo, M., Jain, A., Zhang, H.J.: Content-based hierarchical classification of vacation images. IEEE International Conference on Multimedia Computing and Systems, vol. 1, pp. 518–523 (1999)
41. Vailaya, A., Jain, A., Zhang, H.J.: On image classification: city images vs. landscapes. Pattern Recognit. **31**(12), 1921–1935 (1998)
42. Vailaya, A., Figueiredo, M.A., Jain, A.K., Zhang, H.J.: Image classification for content-based indexing. IEEE Trans. Image Process. **10**(1), 117–130 (2001)
43. Vasconcelos, N., Lippman, A.: Library-based coding: a representation for efficient video compression and retrieval. In: Data Compression Conference, 1997. DCC'97. Proceedings. IEEE (1997), pp. 121–130
44. Vogel, J., Schiele, B.: Semantic modeling of natural scenes for content-based image retrieval. Int. J. Comput. Vis. **72**(2), 133–157 (2007)
45. Wolpert, D.H.: Stacked generalization. Neural Netw. **5**(2), 241–259 (1992)
46. Wu, J., Rehg, J.M.: Centrist: a visual descriptor for scene categorization. IEEE Trans. Pattern Anal. Mach. Intell. **33**(8), 1489–1501 (2011)
47. Xiao, J., Hays, J., Ehinger, K.A., Oliva, A., Torralba, A.: Sun database: large-scale scene recognition from abbey to zoo. In: IEEE Conference on Computer Vision and Pattern Recognition (CVPR), 2010. IEEE (2010), pp. 3485–3492
48. Zhang, L., Li, M., Zhang, H.J.: Boosting image orientation detection with indoor vs. outdoor classification. In: Sixth IEEE Workshop on Applications of Computer Vision, 2002. (WACV 2002). Proceedings. IEEE (2002), pp. 95–99

Chapter 4
Outdoor Scene Classification Using Labeled Segments

Keywords Big visual data · Outdoor scene classification · Image segmentation · Contour-guided color palette · Segmental labeling · Coarse semantic segmentation · Structured machine learning system

4.1 Introduction

In the previous chapter, we discussed low-level feature based indoor/outdoor classification algorithms and meaningful combinations with data grouping and decisions stacking techniques. Since indoor images and outdoor images are quite different in general, it is natural to make computer to treat them separately. As the proposed EDF indoor/outdoor classification system can achieve a quite accurate classification performance robustly, we can move the to second scene understanding stage after the indoor/outdoor classification.

Extending the previous chapter, in this chapter, we will discuss about the semantic-feature based outdoor scene classification using robust segmentations. Outdoor scene classification is a high-level image understanding problem that targets at classifying an outdoor image into one of a number of semantic scene categories (e.g., coast, inside city and street.). It finds applications in image retrieval, 3D layout reconstruction, and context-based object detection, among many others. Different from the previous chapter, we explore the strength of semantic content as image descriptors in the general outdoor scene classification problem. Traditionally, outdoor scene classification is achieved by extracting features from pixels, super-pixels or blocks of an image to characterize its local and global visual patterns and, then, a learning algorithm is adopted to classify the image into a scene category.

In this proposal, we present a new outdoor scene classification method, and call it the coarse semantic segmentation (CSS) method. It is motivated by the observation that most outdoor scene images share many common units such as sky, building, water, plant, mountain, field, road, sand, etc. A proper composition of these units defines a scene. An illustrative example is given in Fig. 4.1, where the input coast image contains four main regions; namely, sky, water, sand and plant. The CSS

C. Chen et al., *Big Visual Data Analysis*,
SpringerBriefs in Signal Processing, DOI 10.1007/978-981-10-0631-9_4

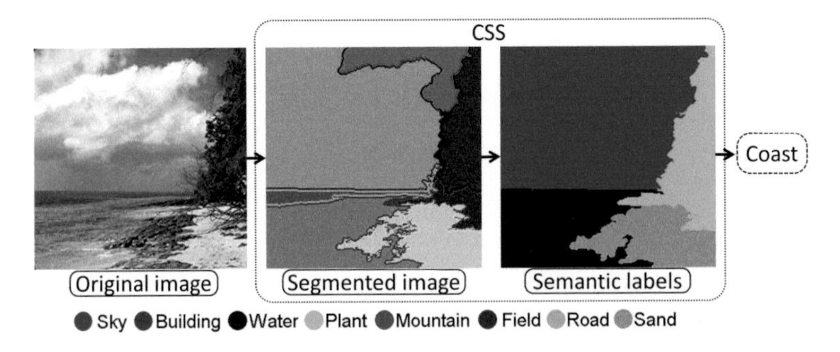

Fig. 4.1 Given an input scene image (*left*), the proposed CSS method conducts a coarse-scale segmentation (*middle*) and associates each segment with a semantic label (*right*). Then, the co-occurrence, appearance and context features of semantic segments are used for scene classification

method consists of three steps: (1) partition the input image into several large segments; (2) predict the semantic label of each segment; and (3) infer the scene category from labeled semantic segments.

There are several novel contributions in the proposed CSS method. First, it is built upon a coarse segmentation result. Robust image segmentation has been a long standing problem in image analysis. However, a significant progress has been made in recent years, and we will show that good segmentation results can be obtained for outdoor images of our interest. Second, instead of using the *abstract* BoF model for the whole image as the mid-level description, we predict the semantic label of each coarse segment from a set of commonly used semantic vocabularies (e.g., sky, building, and water.) based on low-level features in that segment, and adopt the co-occurrence, visual appearance and geometrical relationship of semantic segments as the mid-level description. Third, there exists spatial dependence between adjacent segments. For example, the sky should be above the water although they share similar colors (blue and white). The spatial relation of adjacent segments can be characterized by a Markov Random Field (MRF) model, which enhances semantic labeling accuracy. Finally, a high-level scene category can be inferred from the mid-level description by a machine learning technique, and a two-stage classification system is designed for this objective along this line. A correct classification rate of 93.51 % is achieved by the proposed CSS method on the SIFTFlow dataset [28, 32], which outperforms all state-of-the-art outdoor scene classification methods.

The rest of this chapter is organized as follows. Related work is reviewed in Sect. 4.2. Then, the coarse image segmentation step, the segmental semantic labeling step, and the scene classification step of the CSS method are described in Sects. 4.3 and 4.4, respectively. Experimental results are shown in Sect. 4.5. Finally, concluding remarks are given in Sect. 4.6.

4.2 Review of Previous Works

Classification of scene images is a fundamental task in image understanding. Humans can accurately classify hundreds of scene categories while computers struggle with a small number of scene categories with poorer performance [30, 40]. Most previous works on scene classification focus on representing images with low- and mid-level features. Both local [30] and global [32, 39] low-level feature extraction methods have been well studied. As visual patterns of scene images in the same semantic category become more diverse, these methods encounter bottlenecks in their classification performance due to insufficiency in extracting the semantic information from a scene.

4.2.1 Low-Level Features

Several interesting low-level image descriptors have been proposed and widely adopted in early ages. GIST [32] is one of the most typical and significant ones.

GIST

GIST [32] was proposed to describe a scene's global visual pattern according to the defined spatial envelope. Intuitively inspired by the observations in Fig. 4.2, spatial envelope shape is discovered to be very capable in scene recognition. Similar to the vocabulary employed in architecture, in GIST, the Spatial Envelope of an environment is made by a composite set of boundaries, like walls, sections, ground, elevation and slant of the surfaces that define the shape of the space. Interestingly, authors of GIST conduct several subjective test with 17 observers to observe human behaviors on considering several spatial envelope attributes, including naturalness, openness, roughness, expansion and ruggedness. With inspiring observations and analysis of human visual behaving results, GIST was proposed as a image-based representation to simulate low-level visual pattern's spatial envelope properties.

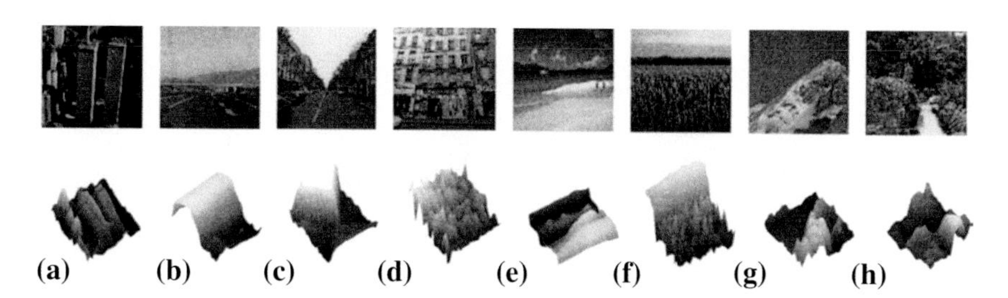

Fig. 4.2 Scenes with different spatial envelopes and their surface representation, where the height level corresponds to the intensity at each pixel (images were low-passed): **a** skyscrapers, **b** an highway, **c** a perspective street, **d** view on a flat building, **e** a beach, **f** a field, **g** a mountain and **e** a forest. The surface shows the information really available after projection of the 3D scene onto the camera. Several aspects of the 3D scene have a direct transposition onto the 2D surface properties (e.g., roughness)

Using the Windowed Fourier Transform (WFT), which is defined as following in Eq. 4.1, GIST tries to describe spatial distribution of spectral information:

$$I(x, y, f_x, f_y) = \sum_{x', y'=0}^{N-1} i(x', y')h_r(x' - x, y' - y)e^{-j2\pi(f_x x' + f_y y')} \tag{4.1}$$

where $h_r(x', y')$ is a hamming window with a circular support of radius r.

To estimate the spatial envelope properties, Discriminant Spectral Templates are used and further extended to Windowed Discriminant Spectral Template (WDST). With thorough spectral-domain analysis, in GIST, scene spatial envelope properties can be evaluated. Figure 4.3 shows an interesting result that organizes scene images according to their two spatial envelope properties, ruggedness and openness.

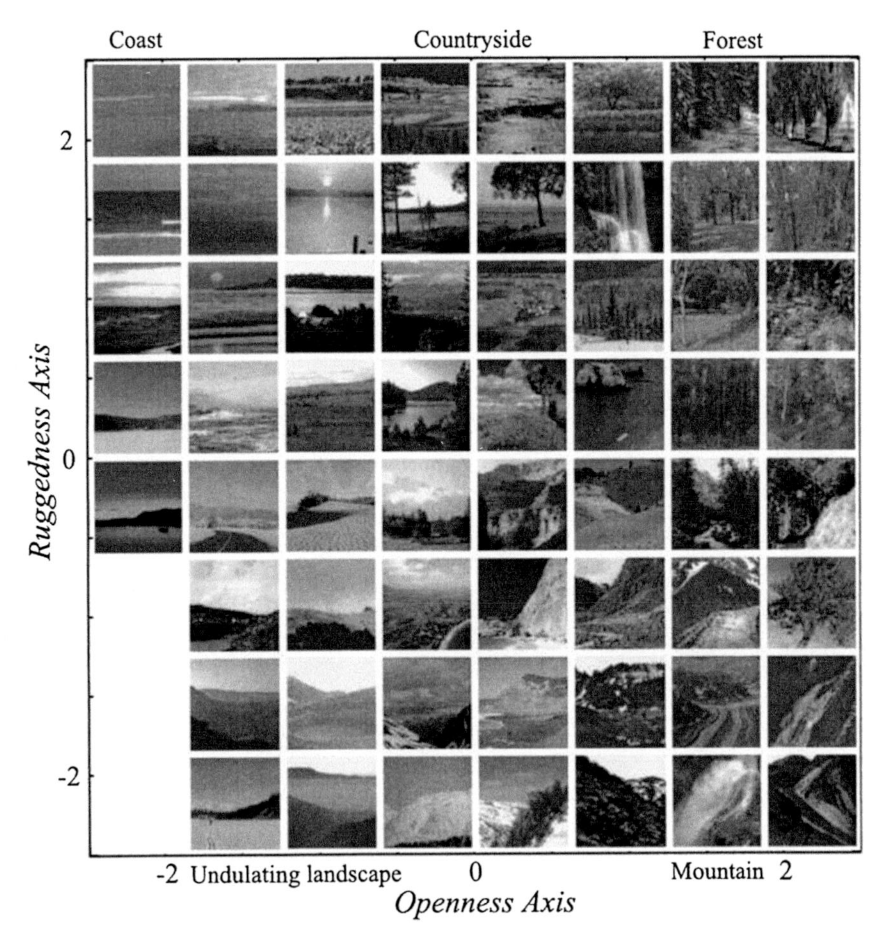

Fig. 4.3 Organization of natural scenes according to openness and ruggedness properties estimated by the WDSTs

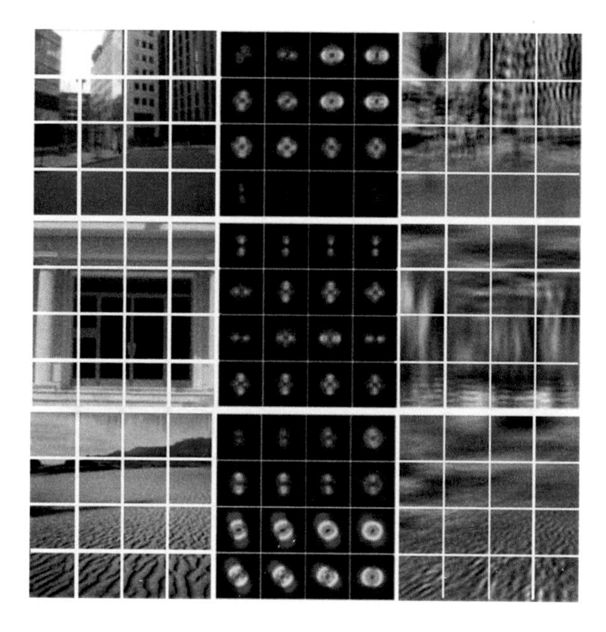

Fig. 4.4 Visualization of Gabor responses in blocks of example scenes. From *left* to *right*: original image; block responses from 32 gabor filters; maximum filter response in gray scale

Unlike the mathematical proof in the paper, on the side of implementation, GIST uses Gabor filters from different scale and orientations to simulate the spectral bases in WDST. With 4 scales and 8 orientations, 32 average Gabor response is calculated in each 1 out of 16 blocks in a image. Catenation of the block responses results in a global texture response of the image. Figure 4.4 shows the block responses in three different image types. We can easily see the differences between scenes and these global image patterns are discriminative in scene categorization.

4.2.2 Mid-Level Features

To describe images in a semantically meaningful way, the mid-level image representation such as the Bag-of-Features (BoF) model [3, 10, 13, 19] uses the frequency of visual words to achieve better performance. The Spatial-Pyramid-Matching (SPM) model [24] and the Oriented Pyramid Matching (OPM) model [41] use the spatial contexts and local orientations of visual words, respectively, to improve classification accuracy furthermore. Besides, the detected objects model [27] and the parts model [21] can be used to capture the semantic content of a scene.

Bag-of-Features (BoF) Models

Bag-of-Feature models were firstly introduced into the scene understanding field by [10]. Borrowing the idea of bag-of-word algorithm in document classification, [10] formed a codebook of key visual words from training images. As shown in Fig. 4.5, the portrait, bicycle and violin have different image patches components. If

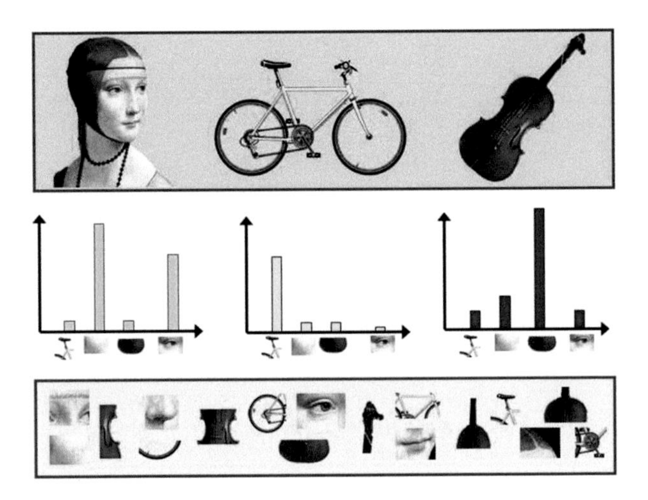

Fig. 4.5 Bag of visual word model in object/scene understanding

we use the bicycle seat, portrait skin, violin bottom and the portrait eye patches as key visual patches, we can construct histograms of the key patches' appearances in the tree images. The histograms can be further used as visual features, which is a mid-level description of the images in classification tasks.

Similarly with examples in Fig. 4.5, the trained visual words and their corresponding mid-level image representation examples of [10] are showed in Fig. 4.6. We can see most visual key words in the trained code book are local patterns with different gradients and orientations. Intuitively, they are treated as the bases of the visual content. By projecting the original into the trained codebook space, [10] is describing the image in a more semantic space than traditional low-level feature based works.

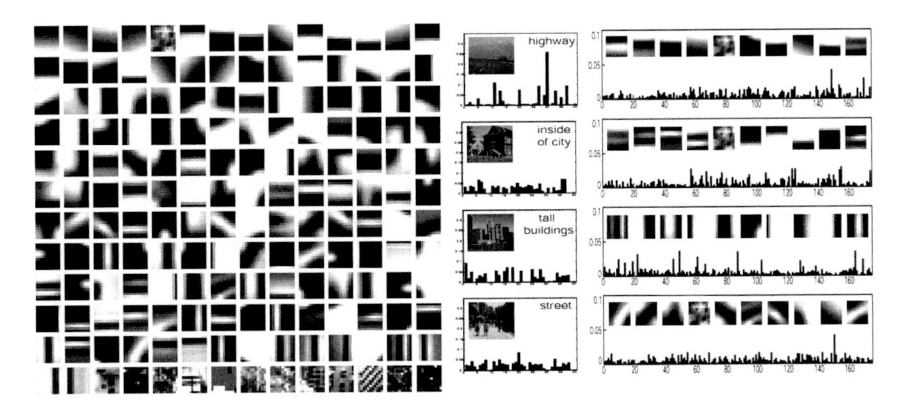

Fig. 4.6 Bag of visual words trained in [10] and corresponding mid-level representations of 4 example scenes

Following [10], many works [3, 13, 19] extend bag-of-word models to bag-of-features models. The visual words in interests are not restricted to pixel patterns (word). More generally, researchers use low-level feature responses, such as SIFT [29] and HOG [7]. Variations of those works lie on codebook training and low-level feature extractions. However, they share the same philosophy and meet performance bottleneck without encoding visual patterns spatial properties. We will see the corresponding improvement from Spatial Pyramid Matching as a extension of the bag-of-features model.

Spatial Pyramid Matching (SPM)

Spatial Pyramid Matching (SPM) [24] aims at encoding spatial descriptions of visual words in an image while extracting BOF features. The original BOF models and its variants extract the appearance frequency of the visual words in an image without considering their spatial information, which is actually quite important in differentiating between detailed scene categories that contain similar content occurrences but different spatial patterns.

As shown in Fig. 4.7, SPM have three level of BOF models. In level 0, histogram of visual words' occurrences is calculated as the original BOF models. In level 1 and level 2, SPM further calculated the similar histograms in sub-regions that are uniformly divided with different scales. In level 1, SPM partitions the whole image into 2×2 subregions. Afterwards, BOF histogram in each sub-region is calculated and cascaded sequentially as the second BOF descriptor. With the concatenation operation, other than the occurrences of the visual codewords, spatial patterns of the visual codewords are also encoded into the descriptor. Similarly but more detailed, in level 2, SPM explores the visual codewords distribution in a even finer scale where image is partitioned into 4×4 grids of blocks. Extractions of cascaded BOF responses in the

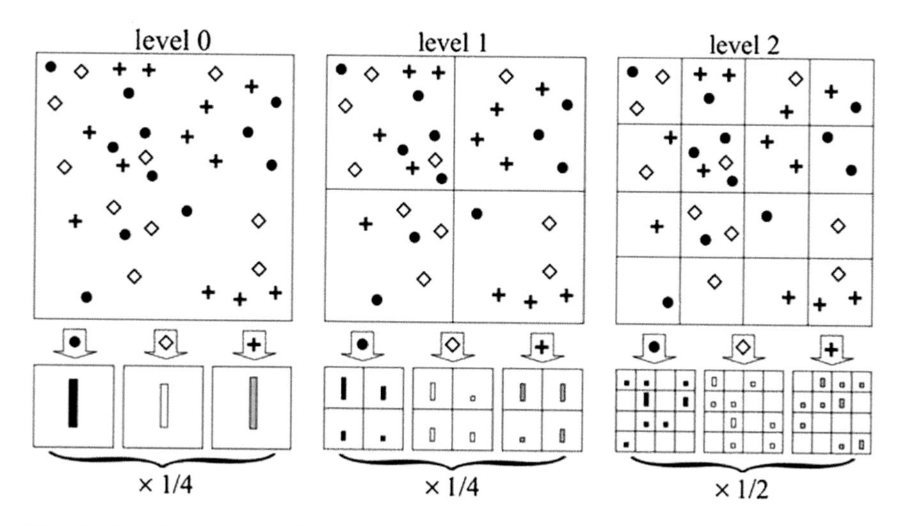

Fig. 4.7 Spatial pyramid matching (SPM) model

three levels results in three histograms which have different numerical scales. To balance the influence from bin values of each histogram in distance measurement, SPM weighted three histograms with different pyramid weight values as shown in Fig. 4.7. As a result, SPM is able to outperform all BOF based approaches and achieve much superior recognition performances in [24]. Conclusively, we observe the importance of semantic visual patterns' spatial distributions from the significant improvement of SPM.

4.2.3 High-Level Features

To further understand the scene concepts with the help of high-level semantic clues such as objects' and surfaces' categories in an image, many recent works start to take advantages of object detection/recognition algorithms. Since object detection recently has achieved breakthroughs with extensive studies and become more accurate and robust in performance. It is natural to apply such techniques to other image understanding tasks like scene understanding. One of the state-of-the-art work that integrates object detectors into scene classification tasks is Object Bank [27].

Object Bank [27] is proposed to represent an image as a scale-invariant response map of a large number of pre-trained generic object detectors, blind to the testing dataset or visual task. Intuitively, from Fig. 4.8, we see that both low-level image descriptors (GIST as an example) and mid-level image representations (SIFT-SPM) are very limited in differentiating these two scenes. In Fig. 4.8, although the mountain scene and the city scene share similar low-level visual patterns (global openness and roughness) and mid-level local visual patterns (SIFT keyword distributions), they have quite different visual concepts such as tree, mountains, building, road etc. Lacking of such high-level recognitions makes computer algorithms impossible to tell the conceptual differences in these two scene images. Correspondingly, Object Bank targets at the solutions to represent scene images with such high-level semantics.

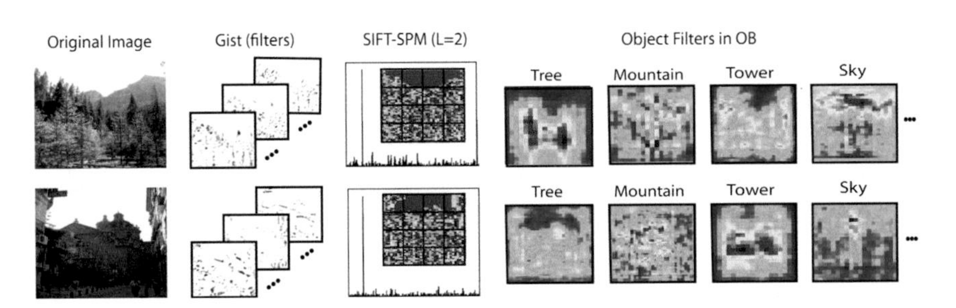

Fig. 4.8 Comparison of object bank (OB) representation with two low-level feature representations, GIST and SIFT-SPM of two types of images, mountain versus city street. From *left* to *right*, for each input image, we show the selected filter responses in the GIST representation, a histogram of the SPM representation of SIFT patches, and a selected number of OB responses

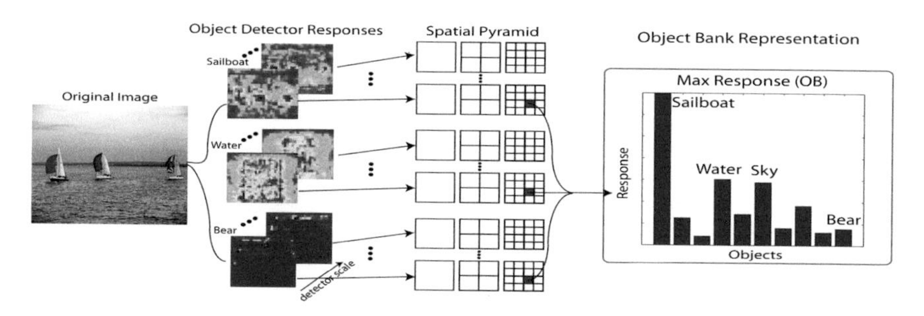

Fig. 4.9 Illustration of OB. A large number of object detectors are first applied to an input image at multiple scales. For each object at each scale, a three-level spatial pyramid representation of the resulting object filter map is used, resulting in $no.objects \times no.scales \times (1^2 + 2^2 + 4^2)$ grids; the maximum response for each object in each grid is then computed, resulting in a no. objects length feature vector for each grid. A concatenation of features in all grids leads to an OB descriptor for the image

Using two state-of-the-art detectors, the latent SVM object detectors [11] and a texture classifier by [20]. Blobby objects such as tables, cars, humans, etc. and texture- and material-based objects such as sky, road, sand, etc. are covered by Object Bank representations. Totally 200 object detections at 12 scales and 3 spatial pyramid levels are used in the OB representations. As shown in Fig. 4.9, each input image will be represented as a cascaded spatially-pooled object distribution histogram using Object Bank. As a result, Object Bank showed impressive performance gains on all traditional scene classification datasets. It proves the significant roles played by conceptual content detections in scene understanding tasks.

4.2.4 *Deep Features*

With the emergence of Convolution Neural Networks (CNN)'s popularities, researchers started to computationally search the best visual feature using CNN framework. Currently, two large image datasets have been used to achieve the goal. They are ImageNet [8] and Places205 [44]. The trained CNN can usually produce a large feature vector as output before the classification stage, and the feature vector is usually called "deep features".

Scene classification using deep features is firstly considered in [30]. Mohan [30] trained and compared two CNN models on ImageNet and Places205 (proposed by [30]). The trained deep features can be visualized in Fig. 4.10. We see the trained feature sets have meaningful patterns in different layer. In the first convolution layer, features are mostly low-level filters. However, in the pooling layers, features usually represent certain local patterns.

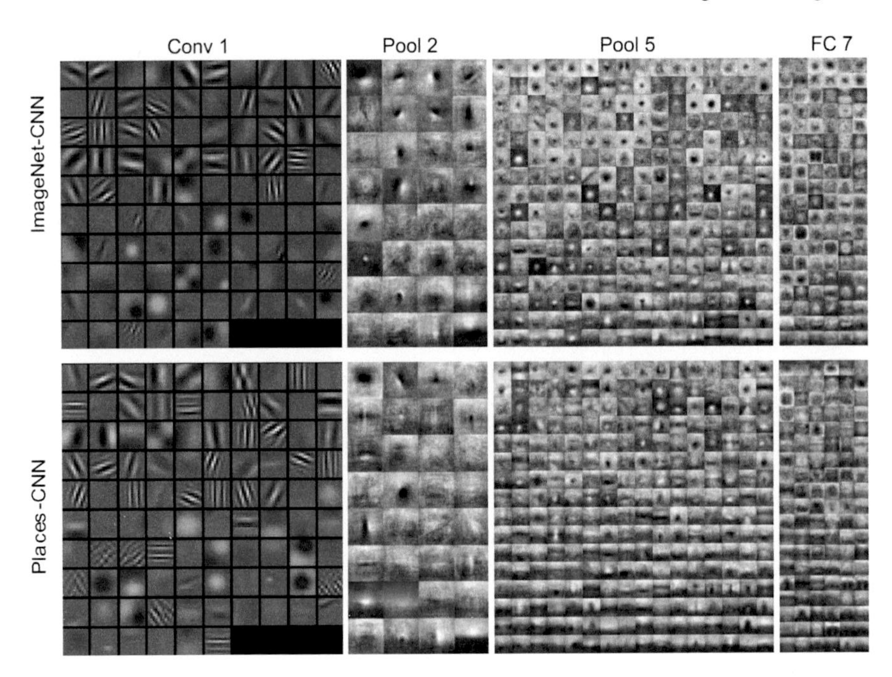

Fig. 4.10 Visualization of the units' receptive fields at different layers for the ImageNet-CNN and Places-CNN. Conv 1 units contains 96 filters. The Pool 2 feature map is $13 \times 13 \times 256$; The Pool 5 feature map is $6 \times 6 \times 256$; The FC 7 feature map is 4096×1. Subset of units at each layer are shown

Table 4.1 Classification accuracy/precision on scene-centric databases and object-centric databases for the Places-CNN feature and ImageNet-CNN feature

Scene centric	SUN397	MIT Indoor67	Scene 15
Places-CNN feature	54.32	68.24	90.19
ImageNet-CNN feature	42.61	56.79	84.23
Object centric	Caltech101	Caltech256	Action40
Places-CNN feature	65.18	45.59	42.86
ImageNet-CNN feature	87.22	67.23	54.92

The classifier in all the experiments is a linear SVM with the same parameters for the two features

In [44], ImageNet-CNN and Places-CNN are also applied to traditional dataset to present their superior performance in Table 4.1. In the table, we see that deep features trained using ImageNet and Places205 show different expertise on different datasets. Since Places205 is scene-centric, Places-CNN feature showed superior performance on scene-centric datasets. On the other hand, ImageNet-CNN feature is more powerful when testing data is object-centric. From this observation, we can conclude that deep features are usually extremely data-driven and their strength highly depend on the training data. In scene understanding literature, since we only have ImageNet and Places205 that can be used to train such deep features, and the variation of CNN structure is not much, there is little works focusing on deep feature training. However, the power of such deeply trained features and the capabilities to handle such large data started to draw researchers' attention.

4.2.5 Scene Parsing and Semantic Segmentation

Being closely related to scene classification, scene parsing attempts to detect and segment semantic objects in a scene image simultaneously. A parametric scene parsing approach with graphic models has been well studied. That is, a cost function is defined based on local appearances and semantic contexts between neighboring blocks [5] or super-pixels [14, 25] and minimized to infer optimal labels for each region in an image. A non-parametric approach built upon the idea of image retrieval and matching can also be used to solve the scene parsing problem. For example, the SIFTflow method [28] retrieves the most similar images by pairwise pixel comparison and, then, transfers the ground-truth labels from pixels in retrieved images to those in the query image. The SuperParsing method [36] follows the same idea but uses super-pixels as the basic unit. Recently, several different improvements were adopted to achieve better performance in [9, 15, 35].

All these existing parametric and non-parametric scene parsing methods use pixels or super-pixels as image units for feature extraction and scene classification. However, it is advantageous to adopt segmentation results to solve the scene classification problem. Along this line, a joint semantic segmentation and scene classification framework was considered in [26, 43], where its solution demands a complicated graphic model with a large number of model parameters to determine. As a result, a large number of labeled data are required in the training.

Parametric Scene Parsing Algorithms

Many parametric scene parsing algorithms were proposed in early stages of the research field. The goal is usually set as a holistic scene understanding which decomposes the scene into regions that are semantically labeled and placed relative to each other within a coherent scene geometry by checking their visual patterns with a parametric model. By "parametric", we mean the models are trained with a training dataset, and parameters of the model are finely toned to ensure the optimal performance on the training dataset. Regarding a new input image, without considering the training dataset, the output label is fully decided by the trained parametric model. One of the most commonly used tool to train such parametric model is Conditional Random Field (CRF).

$$
\begin{aligned}
E(\mathbf{R}, \mathbf{S}, \mathbf{G}, \mathbf{A}, v^{hz}, K | I, \theta) &= \mathbf{E_{pos}} + \mathbf{E_{region}} + \mathbf{E_{pair}} + \mathbf{E_{boundary}} \\
&= \theta^{horizon} \psi^{horizon}(v^{hz}) \\
&\quad + \theta^{region} \sum_{r} \psi_{r}^{region}(S_r, G_r, v^{hz}; A_r, P_r) \\
&\quad + \theta^{pair} \sum_{rs} \psi_{rs}^{pair}(S_r, G_r, S_s, G_s; A_r, P_r, A_s, P_s) \\
&\quad + \theta^{boundary} \sum_{pq} \psi_{pq}^{boundary}(R_p, R_q; \alpha_r, \alpha_r)
\end{aligned} \tag{4.2}
$$

As shown in Fig. 4.11, in [14], to achieve the semantic labels as well as segment the images, CRF model's energy functions is defined in Eq. 4.2. Given an image I and model parameters θ, the energy function scores the entire description of the scene: the pixel-to-region associations R; the region semantic class labels S, geometries G, and appearances A; and the location of the horizon v^{hz}. Basically, four terms are considered and minimized simultaneously. They are position cost ($\mathbf{E_{pos}}$), region appearance cost ($\mathbf{E_{region}}$), pairwise inter-region potential cost ($\mathbf{E_{pair}}$) and a contrast-dependent pairwise boundary potential cost ($\mathbf{E_{boundary}}$). The first two terms evaluate the visual patterns of observed local regions and the last two terms constrain

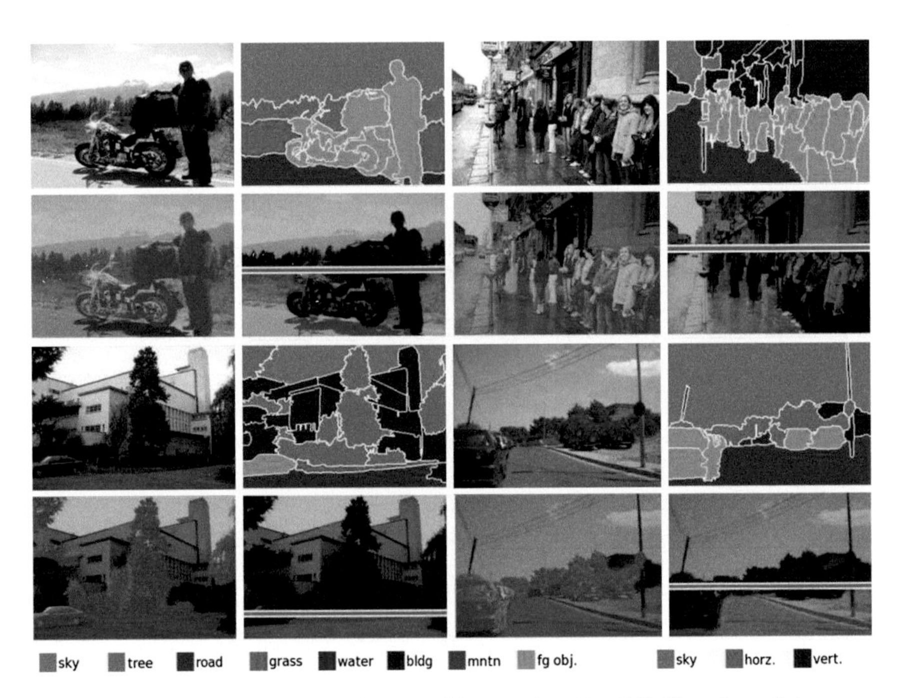

Fig. 4.11 Examples of typical scene decompositions produced by [14]. Show for each image are regions (*top right*), semantic class overlay (*bottom left*), and surface geometry with horizon (*bottom right*)

neighboring consistencies. Fine-toned parameters in such a energy function are achieved on all the training data by standard cost minimization algorithms and finally applied to inferences of testing samples. Such typical parametric approaches can obtain efficient semantic labeling models with finite parameters. It shows certain strength when data diversity is limited and target semantic labels are not many as shown in Fig. 4.11. However, the generality and scalability of the models are bottlenecked by definitions of local regions and global energy functions. Besides, visual pattern diversities in training dataset might not be well preserved into the finite parameters in the models.

Non-parametric Scene Parsing Algorithms

Unlike parametric models, researchers started to explore the data-dependence approach in the scene parsing researchers. One of the pioneers and state-of-the-art algorithms is SIFTFlow [28]. Instead of training parametric discriminative models using training dataset, SIFTFlow take the advantages of the full training data set and segmentally label the target image in a context-preserved label transfer fashion.

As shown in Fig. 4.12, for a given query image, pairwise comparisons on visual patterns is conducted first in SIFTFlow. GIST feature is used for the first-round matching. As a result, a group of scene-level similar images are matched from the training samples. At the second matching stage, a local-scale dense SIFT feature (SIFTFlow) is used to further refine the match result in the retrieved image set and narrow down the most similar image to 3 nearest neighbors in SIFTFlow feature

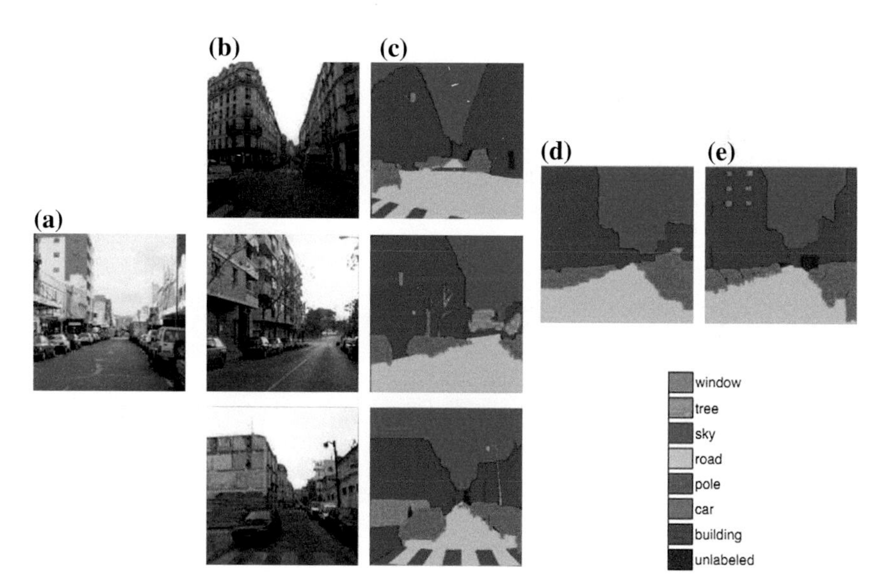

Fig. 4.12 For a query image (**a**), SIFTFlow system finds the top matches (**b**) (three are shown here) using a modified, coarse-to-fine SIFT flow matching algorithm. The annotations of the top matches (**c**) are transferred and integrated to parse the input image as shown in (**d**). For comparison, the ground-truth user annotation of (**a**) is shown in (**e**)

domain. These 3 most similar images are desired to contain the most similar surfaces with the query image's among all the training images. After the matching, label transfer takes place by aligning the pixels in the query image with the retrieved 3 images' that have been semantically labeled. To constrain the spatial consistency between neighboring pixels, a probabilistic Markov Random Field (MRF) model is built. In the final results, as the example in Fig. 4.12, accurate semantic segmentation can be achieved as long as the retrieved training samples are highly relevant to the query image and contains all the semantic components in the query image.

As shown in Fig. 4.13, similar to SIFTFlow [28], SuperParsing [36] extends the non-parametric approach using super-pixels as learning units. Since super-pixels are able to preserve more global visual patterns in local regions than pixels, using super-pixels will create more semantic-level advantages. As a result SuperParsing observes huge performance gain with much more efficient computations than SIFTFlow. Many latter works [9, 15, 35] follow this direction to extend to non-parametric approaches.

Although such non-parametric approaches achieve state-of-the-art results in existing dataset, they usually require robust and accurate image retrieval results, which is not always possible as data patterns diverse with increasing sizes of dataset. Besides, most of them have exhaustive pairwise comparisons which are computationally heavy for large-scale scene understanding. These drawbacks of non-parametric approaches still keep the important role of parametric approaches in the scene parsing field. It is hard to claim which approach is superior in general.

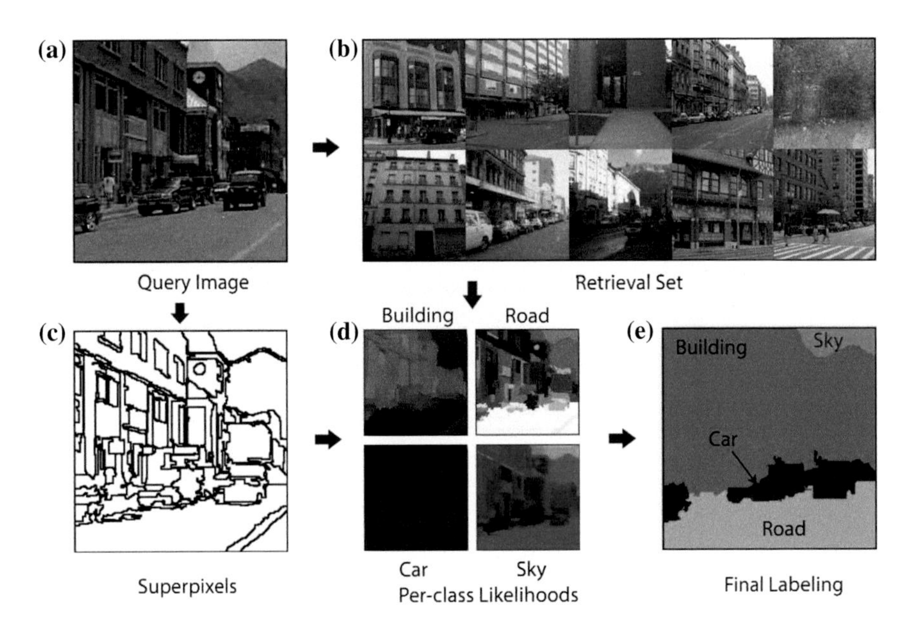

Fig. 4.13 Given a query image (**a**) SuperParsing retrieves similar images from the training dataset (**b**) using several global features. Next, SuperParsing divides the query into super-pixels (**c**) and compute a per-superpixel likelihood ratio score for each class (**d**) based on nearest-neighbor super-pixel matches from the retrieval set. These scores, in combination with a contextual MRF model, give a dense labeling of the query image (**e**)

In conclusion to the scene parsing field, although researchers have spent much effort in this field, two common issues are still observable. On one hand, all of them use pixels, patches and super-pixels as basic learning unit in the segmental labeling stage. Indeed, we know such local-restricted units contain little global information which actually is very significant when human observe and understand visual concepts in scene images. On the other hand, none of them consider adopting the semantic labels achieved to more general scene concept understanding tasks such as scene classification. Correspondingly in this work, we will study the two insufficiencies observed in these traditional scene parsing works and propose solutions to improve scene understanding performance.

4.3 Proposed Coarse Semantic Segmentation (CSS)

As we introduced in the sections above, all traditional scene understanding algorithms extract low-level/mid-level/high-level features from pixels, super-pixels or blocks of a fixed size. However, these local learning units lack of global-scale visual description generalities which is extreme important to scene understanding tasks especially for outdoor scenes where semantical large surfaces play significant roles.

4.3.1 Limitations of Traditional Learning Units

In most traditional scene understanding algorithms [22, 32, 39], image blocks are the first choices as the feature extraction units. However in Fig. 4.14, we can simply find blocks' limits in describing discriminative surface properties with limited local scales. In the figure, three visually typical blocks are shown to appear in both two scene images. It is obvious that blocks cannot be used to extract visual features to differentiate the two scenes.

There are also many works [3, 10, 14, 25] take advantage of super-pixels, which are often generated by MeanShift (MS) or FH [12]. Although these units are simple to obtain, they are *not* natural composition units of a scene image.

Three image composition units (i.e., blocks, super-pixels and segments obtained by the CSS method) are shown in Fig. 4.15. Apparently, like blocks, super-pixels are also not well associated with the semantic decomposition of a scene image because of lacking robustness in segmentation scale and boundary accuracy. In contrast, a good segmentation result will provide a strong correlation between decomposed segments and semantic units of the scene image as shown in Fig. 4.15. The segmentation-based scene analysis methodology has not yet been popular due to the lack of a satisfactory scene image segmentation algorithm. We will show how to overcome this barrier in this section.

Fig. 4.14 Illustration of block limits in describing distinctive parts in two scenes

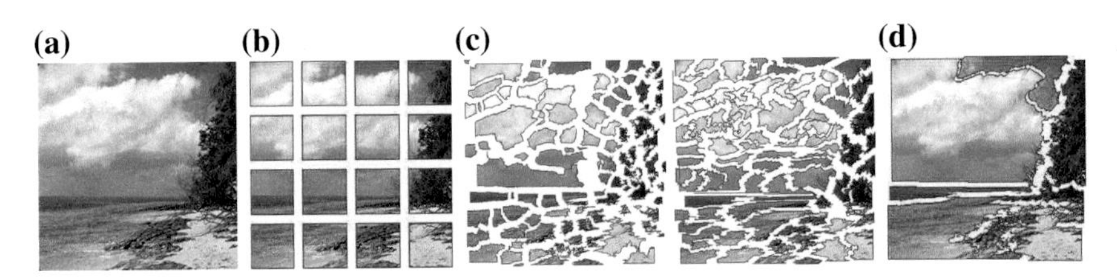

Fig. 4.15 Comparison of feature extraction units. **a** Original. **b** Blocks. **c** MS and FH [12]. **d** CSS

4.3.2 Coarse Segmentation

In Fig. 4.16, we present the proposed robust and image-adaptive coarse segmentation tool in the CSS system. The basic idea of the proposed coarse segmentation method can be simply stated as follows. In order to achieve meaningful segments as basic learning units, as shown in Fig. 4.15, we extract main color components to represent the image content. To further improve the accuracy and significance of the extracted color components and to ensure reasonable spatial connections between neighboring pixels, we also include the main contours of a scene image in the pre-processing and post-processing steps. The flow chart and the corresponding results of each step are illustrated in Fig. 4.16.

Pre-processing. To obtain coarse segments, we focus on pixels along important contours between large homogeneous regions. To achieve this objective, we first denoise the input image with bilateral filtering to reduce color variation caused by noise and texture (e.g., sands and waves as shown in Fig. 4.16). Next, we use the

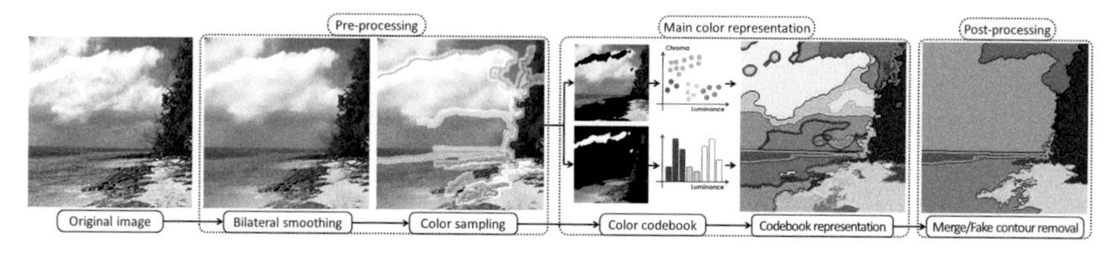

Fig. 4.16 Flow chart of the proposed coarse segmentation in CSS

Canny edge detector to find edge points, where a fixed Gaussian smoothing parameter $\sigma = \sqrt{2}$ and image histogram-adapted hysteresis thresholds are adopted, and link these detected edge points into contours. Lengths of detected contours are considered so as to filter out short contours in textured regions. Afterwards, We define a banded region of interest (ROI) along long contours and collect color samples from pixels in this region by uniform sampling.

Color Codebook Generation. With the collected color samples, we conduct color quantization to reduce the number of representative colors furthermore. We first transform color samples from the RGB to the LCH color space, divide them into "colorful" and "colorless" samples, and treat them differently. That is, samples with a large chroma value, called colorful ones as shown in the first row of the color codebook stage in Fig. 4.16, can be reliably quantized in the Hue channel while colorless samples whose chroma value is small as shown in the bottom row of the color codebook stage in Fig. 4.16 can be quantized in the Luminance channel. We choose an empirical threshold, $T_c = 0.2$, in the Chroma channel to separate colorful and colorless samples for outdoor scenes, where there is less variation in the lighting condition. The MS algorithm with fixed bandwidths, $B_H = 5$ and $B_L = 5$, is applied to colorful and colorless samples, respectively. This process results in an image-adaptive color codebook for input image quantization.

Post-processing. Two post-processing techniques are developed to enhance the quantized scene images furthermore. They are fake boundary removal and small enclosed region mergence. The fake contours in the sky region as shown in Fig. 4.16 are caused by color quantization and should be removed. For their removal, we check whether there are obvious edges along newly generated contours. If no, they are removed. Furthermore, small isolated segments such as the bright cloud and wave segments in Fig. 4.16 are not desired in coarse segmentation since they do contribute little to the scene understanding tasks. They are merged to their surrounding large regions by considering their sizes and ratios of their perimeters to those of their neighbors. The post-processed result of the given example is given in the right-most image in Fig. 4.16.

As we can see more examples in Fig. 4.17, based on such extensive experiments on the 8-scene image dataset, it is worthwhile to emphasize that this module can offer reliable segmentation results for outdoor scene images without the need of parameter fine-tuning.

Fig. 4.17 More coarse segmentation results in CSS: each row contains 2 groups of results. In each group, from *left* to *right* original images with color sampling band highlighted; color quantization results before post-processing; final segmentation result with averaged color segments; final segmentation result overlapped with segmentation contour

4.3.3 Segmental Semantic Labeling

After the coarse image segmentation module, we extract features from each segment and assign a semantic label to it in the second module known as the segmental semantic labeling module. In particular, we select eight target labels which appear frequently in outdoor scene images; namely, sky, water, sand, mountain, field, plant, building and road. There are three steps in this stage as shown in Fig. 4.18, and they will be detailed below.

Feature Extraction. Four types of low-level features are extracted from each segement in CSS—its location, color, texture and line orientation. For segment's position feature (denoted by Pos), we partition an image into $8 \times 8 = 64$ subregions, each of which defines a bin, and calculate the histogram of all pixels in the segment over these 64 bins. For segment's color feature, we adopt the 256-bin color histogram using the standard 256-color pallets, which has 8 uniformly quantized bins in the red and green channels, and 4 bins in the blue channel. For segment's texture feature, we consider the averaged Gabor filter responses (denoted by AvgGabor) at 8 orientations and 4 scales. To achieve the orientated responses in different resolution scales, we follow the GIST [32] in response calculation except that the response is averaged over the segment instead of image sub-blocks. For segment's line orientation (denoted by LineOri), we apply the Canny detector and then calculate a 16-bin orientation histogram of detected edges in the segment.

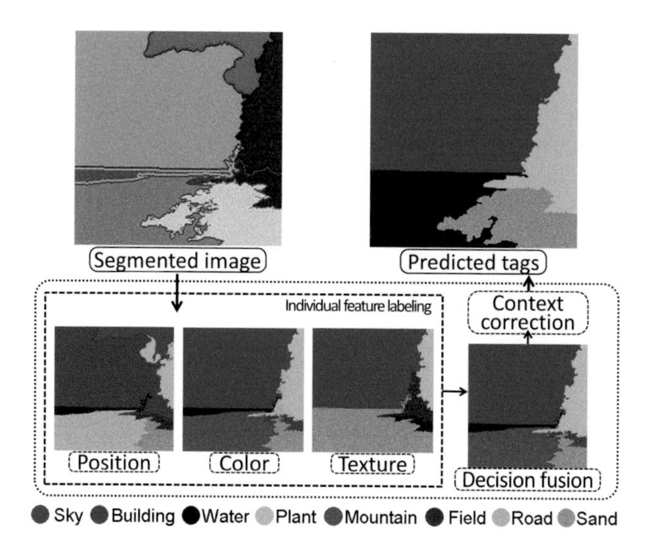

Fig. 4.18 Semantic labeling in CSS

Decision fusion. After extracting low-level features, we train one-versus-all SVM classifiers with different low-level features. At the decision fusion stage, a stacking classifier is trained to fuse the capabilities of the proposed features. As the inputs to the decision fusion classifier, soft decision scores from different features can be achieved with the trained SVM classifiers respectively for each segment. Such soft scores indicate the possibilities that a segment belongs to one of the eight semantic tags according to the corresponding features. We then normalize the soft scores to the range [0, 1] as the confidence of semantic labeling from each feature. As shown in Fig. 4.18, we see the most confident labeling results from different features. Diversities and limitations of individual features can be easily observed. In Fig. 4.18, we find the "sky" segment is falsely labeled as "building" by Pos, since there are many "building" segments in the training data that locate at the tops of images. However, in the RGB, AvgGabor and LineOri's result, we have more accurate decisions for the "sky" segment be cause of its discriminative color and texture appearances. To robustly fuse the opinions from different features, we simply cascade the labeling confidences from each feature as a decision feature and train a meta-level decision stacking SVM classifier.

Context correction. Many state-of-the-art semantic segmentation approaches adopt graphical models, such as Conditional Random Field (CRF) [14] and MRF [28, 36], to achieve contextual inference of neighboring pixels or super-pixels. Considering the advantages from our robust coarse segmentation in CSS, we simply adopt the MRF-based approach to further consider the contexts of coarse segments. Similarly with [36], we define a standard MRF energy function in Eq. 4.3 over the coarse segments' semantic labels $\mathbf{L} = \{l_i\}$ achieved from the decision fusion stage:

$$E(\mathbf{L}) = \sum_{n_i \in N} E_{data}(n_i, l_i) + \lambda \sum_{(n_i, n_j) \in M} E_{smooth}(l_i, l_j) \quad (4.3)$$

where N is the segment set achieved, M is the set of adjacent segment pairs and λ is the parameter to adjust the significance of smoothness constraints. The data term is defined as $E_{data}(n_i, l_i) = 1 - C(n_i, l_i)$, where $C(n_i, l_i)$ is the confidence that segment n_i is predicted with label l_i. The smoothness term is trained based on the occurrence frequency of adjacent segments' labels in all training images:

$$\underset{l_i \neq l_j}{E_{smooth}(l_i, l_j)} = 1 - [\frac{1}{Z_i} F(l_i, l_j) + \frac{1}{Z_j} F(l_j, l_i)] \quad (4.4)$$

where $F(l_i, l_j)$ indicates the number of occurrences that one segment with ground-truth label l_i has a neighbor with label l_j. Z_i is the total counts of adjacent segment pairs, serving as a normalization term. With the defined MRF energy function in Eq. 4.3, we perform MRF inference by minimizing the cost for each image, using a graph cut-based energy minimization toolbox [6, 23, 38]. Observed from results in our dataset, in Figs. 4.18 and 4.19, Eq. 4.4 can provide plausible contextual penalties, especially when improbable adjacent label pairs, such as "water" and 'road", "building" and "sand" appear in the decision fusion results.

Fig. 4.19 CSS visual results *left image* in each column is the segmentation contours in original images. *Right image* in each column is the semantic labels achieved with CSS

4.4 Scene Classification Using CSS

With the semantic segmentation results from CSS, scene classification task can be easily completed by describing the tag's co-occurrence, their contexts and visual appearances in an image. To achieve meaningful descriptions, we design several features accordingly. In addition, similarly with the proposed semantic labeling method in CSS, we also consider a decision fusion stage to combine the strengths from different features.

Tag co-occurrence features. Semantic segments' sizes (TagSize) are computed by evaluating the co-existence and normalized semantic segments sizes. If a tag does not appear in an image, we set the corresponding tag's size to 0. Otherwise, we count the number of pixels in a semantically connected segments and normalize it according to the image size. With the robust CSS results, TagSize can reliably describe the semantic ingredients across different images.

Contexts features. To further consider the spatial contexts of semantic tags in an image, NeighborTag is computed by checking the corresponding tags of upper and lower connections of each segment. In addition, ContentGrid is used to check the tags' spatial distribution over a grid mask in an image. Inheriting the advantage we achieved from large segments and accurate semantic labels in CSS, NeighborTag and ContentGrid are able to efficiently represent the semantic segments layout for the purpose of scene categorization.

Appearance features. We define Vspan as the vertical range of a semantic tag in a image to differentiate appearance of the same semantic surface in different scene images. For example, "inside city" and "tall building" scenes usually have close shots of building facades and far shots of sky scrappers respectively. Comparing to "street" scenes, "inside city" and "tall building" have larger Vspan values for the "building" segments. In this case, Vspan can be highly determinative as an appearance feature. Similarly, TagGabor is designed to describe the texture patterns of each semantic segments in an image. Further differentiations between scenes such as "inside city" and "tall building" require comparing the texture patterns on the "building" surface. In "inside city" scenes, close shot of stores or apartment buildings normally have large grid-like windows. However, in "tall building" scenes, "building" surfaces usually have small and condensed windows. With the proposed TagGabor feature, we can easily differentiate the two scenes which have very similar content ingredients and layouts but quite different semantic surface appearances.

Training and testing. In the training process, we use the first half of the training images (T1) to train individual feature-based SVM classifiers. Then we achieve the categorization soft scores from individual features on the other half of the training set (validation set T2). By cascading the soft scores from all features, we can train a meta-level SVM stacking classifier using T2. Similarly, in the testing stage, we firstly predict soft classification scores of an testing image. Then cascaded decisions from each classifiers are used as decision features for the stacking classifier's input. In the training process, features are extracted using the ground-truth semantic tags provided by human. However, in testing process, we use the predicted semantic labeling results from the proposed CSS. As a conclusion, in the implemented CSS-based classification system, the only difference lies in the usage of human labeled tags and predicted semantic tags in training and testing stages respectively.

4.5 Experimental Results

4.5.1 Dataset

We evaluate the proposed CSS-based outdoor scene classification system on a dataset composed of 2,688 images from 8 outdoor scene categories. This dataset was released for the purpose of scene classification in [32] and later throughly labeled by LabelMe [34] users for the purpose of scene parsing research. This dataset is known as "SIFT Flow dataset (SIFTFlow)" in scene parsingworks [9, 15, 28, 36, 37, 42]. To our knowledge, SIFTFlow is the largest outdoor scene classification dataset with complete semantic labels. In total, it has 33 semantic labels from small foreground objects such as "bird" and "sign" to large background surfaces such as "sky" and "building". Since we only consider large background surfaces in an image, we remove several small object tags. Furthermore, we also merge some visually and functionally similar labels such as "sea" and "water", "grass" and "field", "plant" and "tree". Finally, we have 8 significant large surface labels, "sky", "building", "water", "plant", "mountain", "field", "road" and "sand". In our experiments, following the traditional evaluation formulation, 2,688 images are randomly divided into 1,600 training images (100 training data, 100 validation data from each category) and 1087 testing images in each split. We average the testing performance of such 10 randomly generated splits to compare the CSS-based outdoor scene classification system with previous works.

4.5.2 CSS

As shown in Table 4.2, the conventional Per-Pixel labeling Accuracy (PPA) metric is used in evaluating the performance of CSS. Different features achieve various performances on different semantic tags. For example, Pos works robustly for semantic surfaces like "sky" and "road", which have relatively consistent geometry positions in all images in the SIFTFlow dataset. AvgGabor works well on semantic surfaces such as "building" and "plant", which usually have distinguishable texture patterns across different images. It claims the best individual features among all the proposed low-level features. With the decision fusion scheme, we further achieve nearly 11 % overall accuracy improvement over AvgGabor. In Table 4.2, we can also observe the superior results from MRF-based context correction method. With MRF we achieve

Table 4.2 Performance of CSS's semantic labeling on SIFTFlow with individual features, decision fusion and MRF context correction

	Building (%)	Sand (%)	Field (%)	Mountain (%)	Plant (%)	Road (%)	Water (%)	Sky (%)	PPA (%)
Pos	28.96	11.36	18.86	30.39	32.39	62.43	28.72	79.63	45.28
RGB	46.26	17.90	42.71	33.14	57.69	62.80	24.78	74.50	52.59
AvgGabor	75.25	17.46	50.40	48.37	76.43	63.83	43.55	86.75	68.89
LineOri	73.12	17.97	26.02	43.08	63.41	39.51	38.02	41.53	49.49
Fusion	82.01	12.46	65.33	67.75	76.94	86.32	68.45	93.79	79.43
MRF	89.94	9.61	63.25	74.14	79.57	86.20	72.95	93.73	82.51

an additional 3 % improvement as the final semantic tagging results. Overall, the proposed CSS system achieves a per-pixel accurate labeling rate of 82.51 % on the SIFTFlow dataset. In the table, we also find that in the decision fusion and context correction stage, achieve significant improvement in most categories. However, we observe decreased performance using decision fusion in the rare category "sand". This result is also usually observed in other scene parsing works [9, 36]. In the SIFTFlow dataset, "sand" tags are extremely sparse compared with other semantic categories, especially when we have such large and meaningful coarse segmentations using CSS. Besides, different low-level features normally have very diverse decisions over the rare categories. Without enough training samples, neither decision fusion nor MRF context correction can reach reliable discriminations with such diverse decisions from different features over the rare "sand" category. This problem has also been visited by [42] with a rare class expansion approach. In our system, since scene classification features are not sensitive to such minor tagging errors made by the semantic labeling stage in CSS, we do not keep digging in this path. In Fig. 4.19, we present several outputs of CSS for different scene categories. In the figure, we see that CSS can achieve very reliable coarse segmentation results in the left images of each column. Such outstanding segmentation results provide us with significant advantages for the task of semantic labeling. As shown in Fig. 4.19, accurately-labeled semantic regions are achieved without complicated methodologies in the scene classification stage. The achieved results are also quite comparable to the state-of-the-art works [30, 42] in the scene parsing field.

4.5.3 Scene Classification Results

We compare our classification results with existingworks [1–4, 16–18, 31–33, 39, 45] based on pixels, patches and super-pixels, in Table 4.3. The proposed CSS-based scene classification algorithm achieves the best averaged classification accuracy, 93.54 %, which outperforms the state-of-the-art work [33]. Remarkably, in the

Table 4.3 Averaged classification accuracy of CSS-based scene classification comparing to existing works on SIFTFlow dataset

Approaches	Performance (%)
Gupta et al. [17]	82.66
Oliva et al. [32]	83.70
Wu et al. [39]	86.20
Grossberg et al. [16]	86.60
Bosch et al. [3]	86.65
Niu et al. [31]	87.00
Bosch et al. [4]	87.80
Zhou et al. [45]	89.20
Liu et al. [18]	91.30
Bolovinou et al. [2]	91.49
Mayada et al. [1]	91.94
Perina et al. [33]	92.79
CSS (Ours)	**93.54**

proposed algorithm, we do not use complicated graphical models and structured systems. The superior results approve the significant efficiencies and strong capabilities of the proposed coarse segmentation-based semantic labeling methods in CSS.

We also see the corresponding classification confusion matrix, in Fig. 4.20, which is averaged over all 10 splits in the standard experimental setup. We found most categories can be accurately classified with a rate over 90 % except the "open country". Besides, most confusions of the "open country" class come from "coast", "forest" and "mountain". In our error analysis, we find the results are consistence with the semantic confusions between such three cases. In Fig. 4.21, we can see 6 representative "open country" samples are wrongly predicted as "coast", "forest" and "mountain". Observing the right most image in each image group, we see the pro-

	Coast	Forest	Highw.	Inside.	Mount.	Openc.	Street	Tallbld.
Coast	91.9	0.4	3.1	0.0	0.1	4.4	0.0	0.0
Forest	0.0	97.8	0.0	0.0	1.8	0.4	0.0	0.0
Highw.	0.2	0.0	97.0	0.5	0.0	0.7	0.7	1.0
Inside.	0.0	0.0	0.8	92.2	0.0	0.0	3.0	4.0
Mount.	1.3	0.8	0.3	0.0	95.1	2.5	0.0	0.0
Openc.	5.9	2.0	1.2	0.0	2.5	88.3	0.0	0.1
Street	0.0	0.0	2.2	2.8	0.2	0.0	93.9	0.9
Tallbld.	0.1	0.4	0.4	5.6	0.0	0.1	1.2	92.1

Fig. 4.20 Classification confusion matrix on the SIFTFlow database: percentage of testing data in each row category being classified as categories in each column

Fig. 4.21 Examples of wrongly predicted images in the "open country" class. *First row* prediction as "coast"; *second row* prediction as "forest"; *third row* prediction as "Mountain"

posed CSS works well in the sense of semantic segmentation. Although there are still minor labeling errors in the upper-right and lower-left examples, those errors do not have significant impact to scene classification. However, in these images, it is the confusions in category definitions that make the essential challenge to the classification task. Even human beings cannot easily tell the differences between the scene images' categories. These images are actually located at the semantic boundaries between scene categories. With such confusing ground truths in the SIFTFlow dataset, the proposed CSS-based classification algorithm's performance is reasonably bounded.

4.6 Summary

In this chapter, we present a novel CSS-based outdoor scene classification approach. In CSS, we firstly propose a robust segmentation method, which achieves large and reasonably connected segments as the basic learning unit without any image-dependent parameter. Based on the robust segmentation results, we propose an efficient semantic labeling method by combining features with a two-stage stacking system and a MRF-based context-aware model. Finally, we apply the CSS results to the outdoor scene classification problem and achieve the state-of-the-art scene classification performance on a widely used SIFTFlow dataset. Based on a thorough error analysis, we believe the CSS-based approach can achieve even better performance without ground truth confusions in the dataset. Furthermore, we believe the proposed CSS is not restricted to the scene classification problem. It can be applied to any image understanding problems by using the segmentation results in CSS as the basic learning unit.

References

1. Ali, M.M., Fayek, M.B., Hemayed, E.E.: Human-inspired features for natural scene classification. Pattern Recogn. Lett. **34**(13), 1525–1530 (2013)
2. Bolovinou, A., Pratikakis, I., Perantonis, S.: Bag of spatio-visual words for context inference in scene classification. Pattern Recogn. **46**(3), 1039–1053 (2013)
3. Bosch, A., Zisserman, A., Muñoz, X.: Scene classification via plsa. In: Computer Vision-ECCV 2006, pp. 517–530. Springer (2006)
4. Bosch, A., Zisserman, A., Muoz, X.: Scene classification using a hybrid generative/discriminative approach. IEEE Trans. Pattern Anal. Mach. Intell. **30**(4), 712–727 (2008)
5. Boutell, M.R., Luo, J., Brown, C.M.: Scene parsing using region-based generative models. IEEE Trans. Multimedia **9**(1), 136–146 (2007)
6. Boykov, Y., Veksler, O., Zabih, R.: Fast approximate energy minimization via graph cuts. IEEE Trans. Pattern Anal. Mach. Intell. **23**(11), 1222–1239 (2001)
7. Dalal, N., Triggs, B.: Histograms of oriented gradients for human detection. In: IEEE Computer Society Conference on Computer Vision and Pattern Recognition, 2005. CVPR 2005, vol. 1, pp. 886–893. IEEE (2005)
8. Deng, J., Dong, W., Socher, R., Li, L.J., Li, K., Fei-Fei, L.: Imagenet: a large-scale hierarchical image database. In: IEEE Conference on Computer Vision and Pattern Recognition, 2009. CVPR 2009, pp. 248–255. IEEE (2009)
9. Eigen, D., Fergus, R.: Nonparametric image parsing using adaptive neighbor sets. In: 2012 IEEE Conference on Computer Vision and Pattern Recognition (CVPR), pp. 2799–2806. IEEE (2012)
10. Fei-Fei, L., Perona, P.: A bayesian hierarchical model for learning natural scene categories. In: IEEE Computer Society Conference on Computer Vision and Pattern Recognition, 2005. CVPR 2005, vol. 2, pp. 524–531. IEEE (2005)
11. Felzenszwalb, P.F., Girshick, R.B., McAllester, D., Ramanan, D.: Object detection with discriminatively trained part-based models. IEEE Trans. Pattern Anal. Mach. Intell. **32**(9), 1627–1645 (2010)
12. Felzenszwalb, P.F., Huttenlocher, D.P.: Efficient graph-based image segmentation. Int. J. Comput. Vision **59**(2), 167–181 (2004)
13. Gokalp, D., Aksoy, S.: Scene classification using bag-of-regions representations. In: IEEE Conference on Computer Vision and Pattern Recognition, 2007. CVPR'07, pp. 1–8. IEEE (2007)
14. Gould, S., Fulton, R., Koller, D.: Decomposing a scene into geometric and semantically consistent regions. In: 2009 IEEE 12th International Conference on Computer Vision, pp. 1–8. IEEE (2009)
15. Gould, S., Zhang, Y.: Patchmatchgraph: building a graph of dense patch correspondences for label transfer. Computer Vision? ECCV 2012, pp. 439–452. Springer (2012)
16. Grossberg, S., Huang, T.R.: Artscene: a neural system for natural scene classification. J. Vision **9**(4), 6.1–19 (2009)
17. Gupta, D., Singh, A.K., Kumari, D.: Hybrid feature based natural scene classification using neural network. Int. J. Comput. Appl. **41**(16), 48–52 (2012)
18. Han, Y., Liu, G.: A hierarchical gist model embedding multiple biological feasibilities for scene classification. In: ICPR, pp. 3109–3112 (2010)
19. Hao, J., Jie, X.: Improved bags-of-words algorithm for scene recognition. In: 2010 2nd International Conference on Signal Processing Systems (ICSPS), vol. 2, pp. V2–279–V2-282. IEEE (2010)
20. Hoiem, D., Efros, A.A., Hebert, M.: Automatic photo pop-up. ACM Trans. Graph. (TOG) **24**(3), 577–584 (2005)
21. Juneja, M., Vedaldi, A., Jawahar, C., Zisserman, A.: Blocks that shout: distinctive parts for scene classification. In: 2013 IEEE Conference on Computer Vision and Pattern Recognition (CVPR), pp. 923–930. IEEE (2013)

22. Kim, W., Park, J., Kim, C.: A novel method for efficient indoor-outdoor image classification. J. Signal Process. Syst. **61**(3), 251–258 (2010)
23. Kolmogorov, V., Zabin, R.: What energy functions can be minimized via graph cuts? IEEE Trans. Pattern Anal. Mach. Intell. **26**(2), 147–159 (2004)
24. Lazebnik, S., Schmid, C., Ponce, J.: Beyond bags of features: spatial pyramid matching for recognizing natural scene categories. In: 2006 IEEE Computer Society Conference on Computer Vision and Pattern Recognition, vol. 2, pp. 2169–2178. IEEE (2006)
25. Lempitsky, V., Vedaldi, A., Zisserman, A.: Pylon model for semantic segmentation. In: Advances in Neural Information Processing Systems, pp. 1485–1493 (2011)
26. Li, L.J., Socher, R., Fei-Fei, L.: Towards total scene understanding: classification, annotation and segmentation in an automatic framework. In: IEEE Conference on Computer Vision and Pattern Recognition, 2009. CVPR 2009, pp. 2036–2043. IEEE (2009)
27. Li, L.J., Su, H., Fei-Fei, L., Xing, E.P.: Object bank: A high-level image representation for scene classification and semantic feature sparsification. In: Advances in Neural Information Processing Systems, pp. 1378–1386 (2010)
28. Liu, C., Yuen, J., Torralba, A.: Nonparametric scene parsing: Label transfer via dense scene alignment. In: IEEE Conference on Computer Vision and Pattern Recognition, 2009. CVPR 2009, pp. 1972–1979. IEEE (2009)
29. Lowe, D.G.: Distinctive image features from scale-invariant keypoints. Int. J. Comput. Vision **60**(2), 91–110 (2004)
30. Mohan, R.: Deep deconvolutional networks for scene parsing (2014)
31. Niu, Z., Hua, G., Gao, X., Tian, Q.: Context aware topic model for scene recognition. In: 2012 IEEE Conference on Computer Vision and Pattern Recognition (CVPR), pp. 2743–2750. IEEE (2012)
32. Oliva, A., Torralba, A.: Modeling the shape of the scene: a holistic representation of the spatial envelope. Int. J. Comput. Vision **42**(3), 145–175 (2001)
33. Perina, A., Cristani, M., Castellani, U., Murino, V., Jojic, N.: A hybrid generative/discriminative classification framework based on free-energy terms. In: 2009 IEEE 12th International Conference on Computer Vision, pp. 2058–2065. IEEE (2009)
34. Russell, B.C., Torralba, A., Murphy, K.P., Freeman, W.T.: Labelme: a database and web-based tool for image annotation. Int. J. Comput. Vision **77**(1–3), 157–173 (2008)
35. Singh, G., Kosecka, J.: Nonparametric scene parsing with adaptive feature relevance and semantic context. In: 2013 IEEE Conference on Computer Vision and Pattern Recognition (CVPR), pp. 3151–3157. IEEE (2013)
36. Tighe, J., Lazebnik, S.: Superparsing: scalable nonparametric image parsing with superpixels. Computer Vision?ECCV 2010, pp. 352–365. Springer (2010)
37. Tighe, J., Lazebnik, S.: Finding things: Image parsing with regions and per-exemplar detectors. In: 2013 IEEE Conference on Computer Vision and Pattern Recognition (CVPR), pp. 3001–3008. IEEE (2013)
38. Tung, F., Little, J.J.: Collageparsing: Nonparametric scene parsing by adaptive overlapping windows (2014)
39. Wu, J., Rehg, J.M.: Centrist: a visual descriptor for scene categorization. IEEE Trans. Pattern Anal. Mach. Intell. **33**(8), 1489–1501 (2011)
40. Xiao, J., Hays, J., Ehinger, K.A., Oliva, A., Torralba, A.: Sun database: large-scale scene recognition from abbey to zoo. In: 2010 IEEE conference on Computer Vision and Pattern Recognition (CVPR), pp. 3485–3492. IEEE (2010)
41. Xie, L., Wang, J., Guo, B., Zhang, B., Tian, Q.: Orientational pyramid matching for recognizing indoor scenes. In: 2014 IEEE Conference on Computer Vision and Pattern Recognition (CVPR), pp. 3734–3741. IEEE (2014)
42. Yang, J., Price, B., Cohen, S., Yang, M.H.: Context driven scene parsing with attention to rare classes. In: 2014 IEEE Conference on Computer Vision and Pattern Recognition (CVPR), pp. 3294–3301. IEEE (2014)
43. Yao, J., Fidler, S., Urtasun, R.: Describing the scene as a whole: joint object detection, scene classification and semantic segmentation. In: 2012 IEEE Conference on Computer Vision and Pattern Recognition (CVPR), pp. 702–709. IEEE (2012)

44. Zhou, B., Lapedriza, A., Xiao, J., Torralba, A., Oliva, A.: Learning deep features for scene recognition using places database. In: Advances in Neural Information Processing Systems, pp. 487–495 (2014)
45. Zhou, L., Zhou, Z., Hu, D.: Scene classification using a multi-resolution bag-of-features model. Pattern Recogn. **46**(1), 424–433 (2013)

Chapter 5
Global-Attributes Assisted Outdoor Scene Geometric Labeling

Keywords Geometric label · Outdoor scene understanding · Image segmentation · Segmental labeling · Coarse semantic segmentation

5.1 Introduction

Automatic 3D geometric labeling or layout reasoning from a single scene image is one of the most important and challenging problems in scene understanding. It offers the mid-level information for other high-level scene understanding tasks such as 3D world reconstruction [13, 15, 18, 27], depth map estimation [25], scene classification and content-based image retrieval.

Recovering the 3D structure from a single image using a local visual pattern recognition approach was studied in early geometric labeling research [3, 16, 18–21, 23, 25, 31]. To give an example, Hoiem et al. [19] defined seven labels (i.e., sky, support, planar left/right/center, porous and solid) and classified super-pixels to one of these labels according to their local visual appearances. Features such as color, position, texture pattern and segment shape were used to describe local visual properties. Since the same surface (e.g., a building facade) may take a different geometric role in different images, the performance of all local-patch-based labeling methods is limited. To improve the performance of local-patch-based methods, researchers have incorporated the global constraints or context rules in recent years. Gupta et al. [11, 12] proposed a qualitative physical model for outdoor scenes by assuming that objects are composed by blocks of volume and mass. If a scene image fits the underlying 3D model, better surface layout estimation can be achieved. However, their model is not generic enough to cover a wide range of scenes. Liu et al. [26] and Pan et al. [28] focused on images containing building facades and developed multiple global 3D context rules using their distinctive geometry cues such as vanishing lines.

Being motivated by recent trends, we exploit both local and global attributes for layout estimation and propose a Global-attributes Assisted Labeling (GAL) system

C. Chen et al., *Big Visual Data Analysis*,
SpringerBriefs in Signal Processing, DOI 10.1007/978-981-10-0631-9_5

in this work. GAL uses local attributes to provide initial labels for all superpixels and extracts global attributes such as sky lines, ground lines, horizon, vanishing lines, etc. Then, it uses global attributes to confirm or correct initial labels. Our work contributes to this field in two folds. First, it provides a new framework to address the challenging geometric layout labeling problem, and this framework is supported by encouraging results. Second, as compared with previous work, GAL can handle images of more diversified contents using inference from global attributes. The performance of GAL is benchmarked with several state-of-the-art algorithms against a popular outdoor scene layout dataset, and significant performance improvement is observed.

The rest of this paper is organized as follows. The GAL system is described in Sect. 5.3. Experimental results are shown in Sect. 5.4, which includes analysis of several poor results. Finally, concluding remarks are given in Sect. 5.5.

5.2 Review of Previous Works

5.2.1 Geometric Context from a Single Image

Hoiem et al. [19] defined seven geometric labels (i.e., sky, support, planar left/right/center, porous and solid) and classified super-pixels to one of these labels according to their local visual appearances. Features such as color, position, texture pattern and segment shape were used to describe local visual properties. The super-pixel segmentation has two limitations in the geometric labeling problem. Fist, different regions with weak boundaries subject to under-segmentation while texture region subject to over-segmentation. Second, since one label can be assigned to one segment, it could be either a wrong label or a correct label. If the classifier makes a wrong decision, it could not be corrected. One segmentation could not be perfect in geometric labeling work as pointed in [19]. Therefore, the algorithm of a weight sum of decisions from multiple segmentations is proposed in [19]. The flowchat of the proposed algorithm in [19] is shown in Fig. 5.1. Specifically, given the original image, a popular graph-based segmentation algorithm [7] is applied with a larger parameter setting which results in smaller segment unit. Then a superpixel merge algorithm is developed to merge the superpixels into different number of segments. The superpixel merge examples are shown in Fig. 5.2. By weighted sum of the decisions from different segments, the labeling accuracy increase.

The superpixel learning method tries to establish the relation between local superpixel appearance with the desired label, however, due to lack of global information and global physical constraints, the superpixel learning algorithm can give meaningless result.

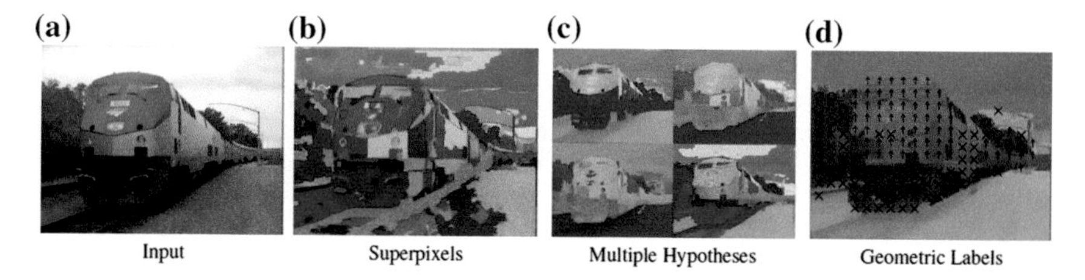

Fig. 5.1 To obtain useful statistics for modeling geometric classes, [19] slowly build structural knowledge of the image: from pixels (**a**), to superpixels (**b**), to multiple potential groupings of superpixels (**c**), to the final geometric labels (**d**)

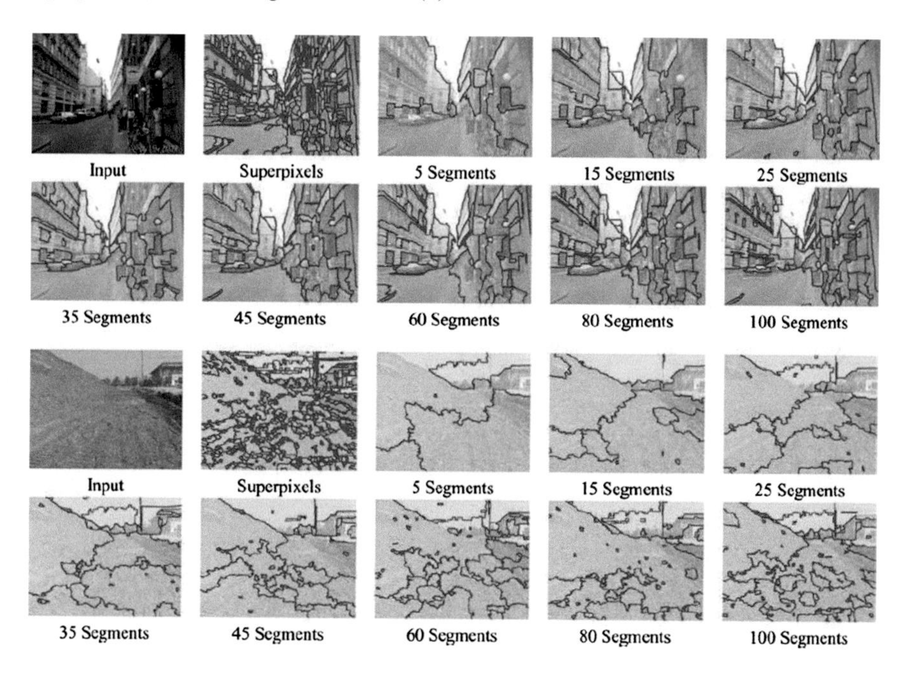

Fig. 5.2 Examples of multiple segmentations [19]

5.2.2 *Blocks World Revisited*

Since the same surface (e.g., a building facade) may take a different geometric role in different images, the performance of all local-patch-based labeling methods is limited. To improve the performance of local-patch-based methods, researchers have incorporated the global constraints or context rules in recent years. Gupta et al. [11, 12] proposed a qualitative physical model for outdoor scenes by assuming that objects are composed by blocks of volume and mass. If a scene image fits the underlying 3D model, better surface layout estimation can be achieved. Figure 5.3 shows eight block-view classes. In the inference process, each segments will be fitted into one of

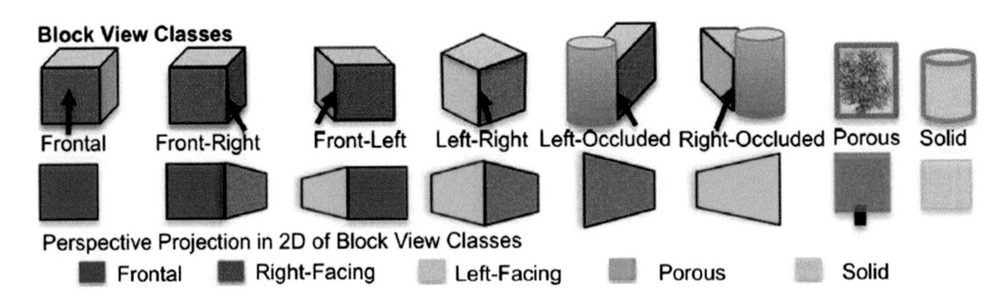

Fig. 5.3 Catalog of the possible block view classes and associated 2D projections. The 3D blocks are shown as cuboids although our representation imposes no such constraints on the 3D shape of the block. The *arrow* represents the camera viewpoint

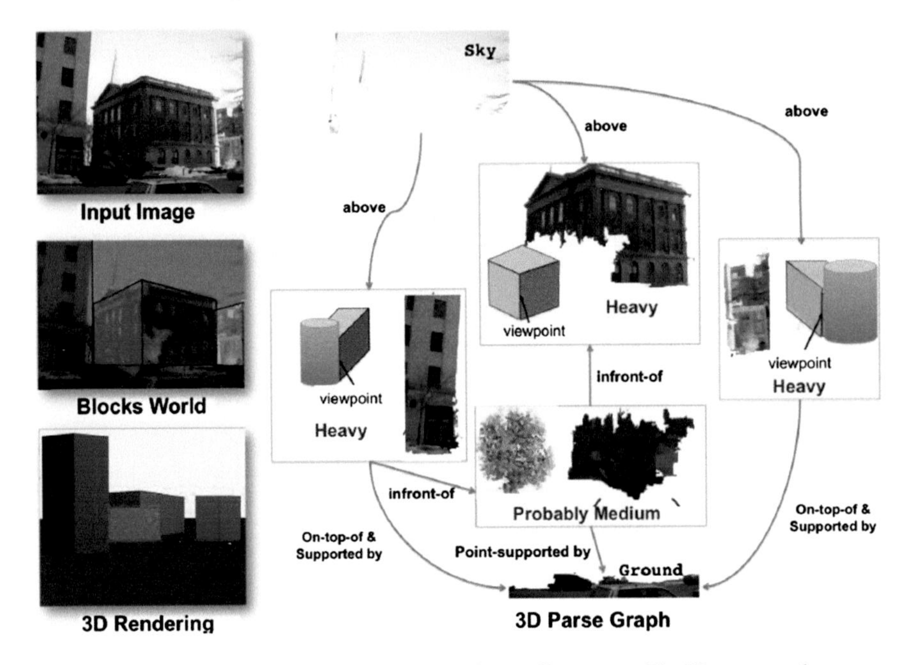

Fig. 5.4 Example output of our automatic scene understanding system. The 3D parse graph summarizes the inferred object properties (physical boundaries, geometric type, and mechanical properties) and relationships between objects within the scene. [11]

the eight block-view classes. Other constraints include geometry constraints obtained from initial label result in [19], contact constraints which measure the agreement of geometric properties with ground and sky contact points, intra-class and stability constraint which measure physical stability within a single block and with respect to other blocks respectively, and finally the depth constraint which measures the agreement of projection of blocks in the 2D image plane with the estimated depth ordering. Given a candidate block B_i, its associated geometrical properties and its relationship to the blocks is estimated by minimizing the following cost function:

$$
\begin{aligned}
C(B_i) = F_{geometry}(G_i) + \sum_{S \in ground, sky} F_{contacts}(G_i, S) + F_{intra}(S_i, G_i, d) \\
+ \sum_{j \in blocks} F_{stability}(G_i, S_{i,j}, B_j) + F_{depth}(G_i, S_{i,j}, D)
\end{aligned}
\tag{5.1}
$$

Figure 5.4 shows one example of parsing graph. By using blocks to fit into a physically world, global information can be added into the geometric labeling. As a result, the wrong local patch labeling error can be avoided. However, the proposed algorithm suffers from limited number of block models which fails to cover all the possibilities in real world. In addition, if the segment is fitted to a wrong block model, the segment will be labeled totally wrong.

5.2.3 Single-View 3D Scene Parsing by Attributed Grammar

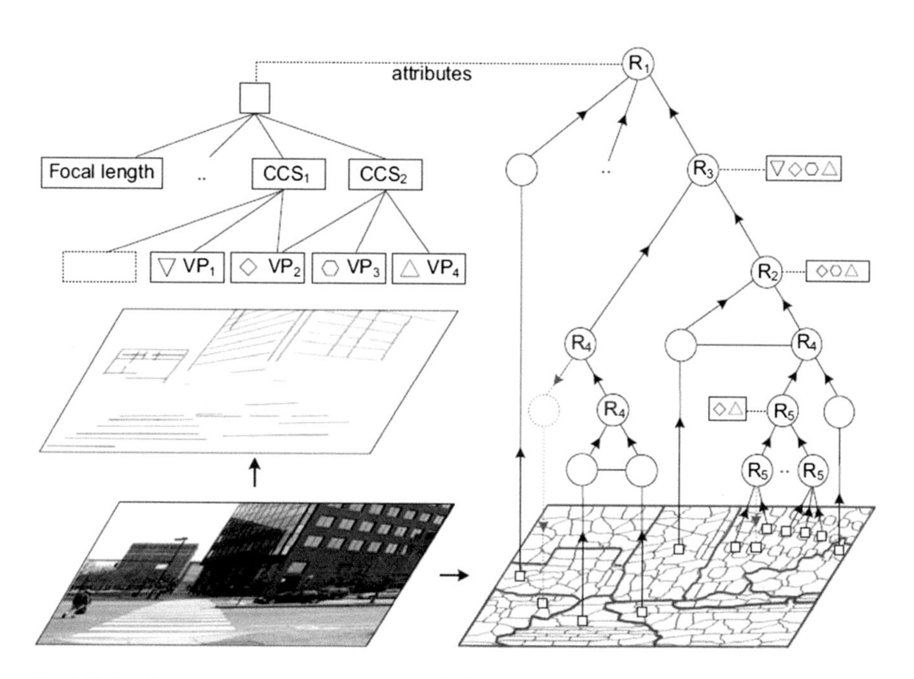

Fig. 5.5 Parsing images using grammar rules [26]

Liu et al. [26] proposed a Bayesian framework and five merge rules to get the geometric label for urban scene. Figures 5.5 and 5.6 show the parsing graph and five grammar rules respectively. The algorithm first find the straight lines and estimates the vanishing points in the image and partition the image into superpixels. The inference

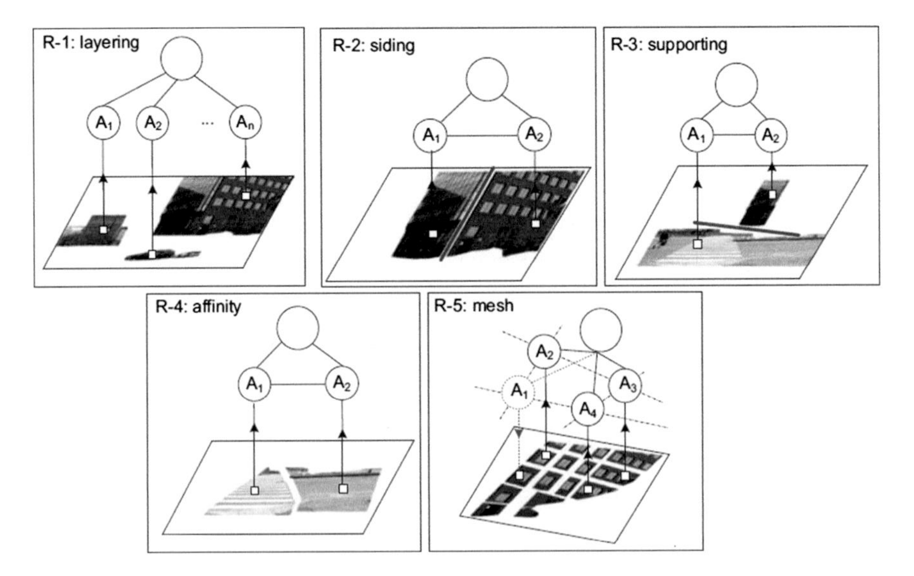

Fig. 5.6 Illustration of the five grammar rules [26]

is based on Composite Cluster Sampling (CCS). The initial covering is obtained by K-means algorithm and then in each iterations, the algorithm makes proposal based on the five grammar rules, namely, layering (rule 1), siding (rule 2), supporting (rule 3), affinity (rule 4) and mesh (rule 5). There are two stages during the inference. In the first stage, the proposals are made using rule 4 and rule 5. In the second stage, the proposals are made from all the five grammar rules.

The algorithm proposed by Liu et al. [26] has the Manhattan world assumption and relies on the an accurate vanishing point detection algorithm. Obviously, the algorithm could not handle most of the nature scenes very well.

5.2.4 *Inferring 3D Layout of Buildings from a Single Image*

Similar with Liu et al. [26]'s work, Pan et al. [28] also focused on images containing building facades. Pan et al. [28] addresses the problem of detecting a set of distinctive facade planes and estimate their 3D orientations and locations given a single 2D image of urban scene. Different from previous methods that only yield coarse orientation labels or qualitative block approximations, the proposed algorithm quantitatively reconstructs building facades in 3D space using a set of planes mutually related by 3D geometric constraints. Each plane is characterized by a continuous orientation vector and a depth distribution. An optimal solution is reached through inter-planar interactions. The proposed model infers the optimal 3D layout of building facades by maximizing a defined objective function. The data term is the product of two scores:

image feature compatibility and geometric compatibility score. The image feature compatibility measures how well the 2D location of a facade plane in the image agrees with image features. The geometric compatibility score measures probability of the ground contact line position. The smoothness constraints include convex-corner constraint, occlusion constraint and alignment constraint.

Pan et al. [28] proposed an algorithm that infers the 3D layout of building facades. Due to the quantitative and plane-based nature of geometric reasoning, the model is more expressive and informative than existing approaches. However, the algorithm does not provide solutions to labeling ground, sky, porous and solid classes which are also important in 3D reconstruction (Fig. 5.7).

5.3 Proposed GAL System

5.3.1 System Overview

The flow chart of the GAL system is given in Fig. 5.8.

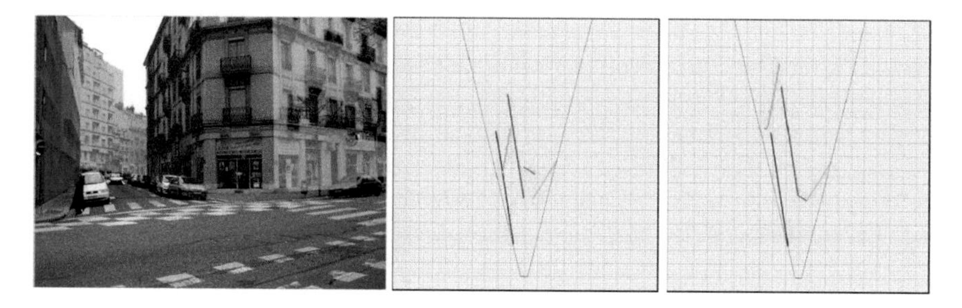

Fig. 5.7 [28] detects building facades in a single 2D image, decomposes them into distinctive planes of different 3D orientations, and infers their optimal depth in 3D space based on cues from individual planes and 3D geometric constraints among them. *Left* Detected facade regions are covered by shades of different colors, each color representing a distinctive facade plane. *Middle/Right* ground contact lines of building facades on the ground plane before/after considering inter-planar geometric constraints. The coarser grid spacing is 10 m

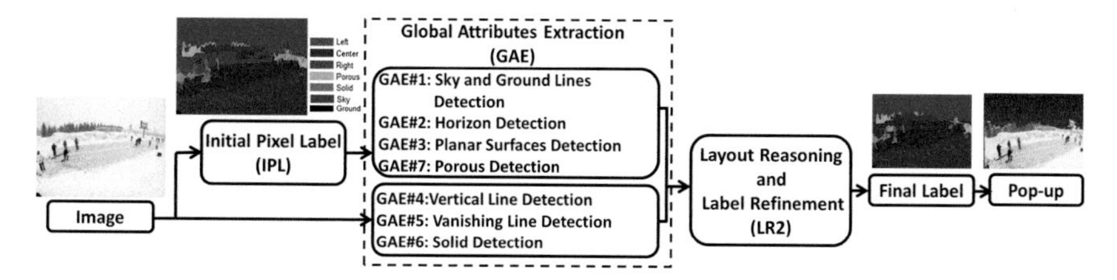

Fig. 5.8 The flow chart of the proposed GAL system

- Stage 1: Initial Pixel Labeling (IPL);
- Stage 2: Global Attributes Extraction (GAE);
- Stage 3: Layout Reasoning and Label Refinement (LR2).

For a given outdoor scene image, we can obtain initial pixel labeling results using any local-features-based labeling method in the first stage. Here, we finetune the classic method by Hoiem et al. [16] slightly for this purpose. The labeling performance of the IPL stage is however not satisfactory due to the lack of global scene information. To address this issue, we pose the following seven questions for each scene image and would like to answer them based on all possible visual cues (e.g., color, edge contour, defocus degree, etc.) in the second stage:

1. Is there sky in the image? If yes, where?
2. Is there ground in the image? If yes, where?
3. Does the image contain a horizon? If yes, where?
4. Are there planar surfaces in the image? If yes, where and what are their orientations?
5. Is there any building in the image? If yes, where and what is its orientation?
6. Is there solid in the image? If yes, where is it?
7. Is there porous in the image? if yes, where is it?

The answers to the first part of each question lead to a 7D global attribute vector (GAV) with binary values (YES or NO), where we set "YES" and "NO" to "1" and "0", respectively. If the value for an entry is "1", we need to provide more detailed description for the corresponding global attribute. The knowledge of the GAV is helpful in providing a robust labeling result. Based on extracted global attributes, we conduct layout reasoning and label refinement in the third stage. Layout reasoning can be greatly simplified based on global attributes. Then, the label of each pixel can be either confirmed or adjusted based on inference. The design and extraction of global attributes in the GAE stage and the layout reasoning and label refinement in the LR2 stage are two novel contributions. They will be elaborated in Sects. 5.3.3 and 5.3.4, respectively.

5.3.2 *Initial Pixel Labeling (IPL)*

The method proposed by Hoiem et al. in [16] offers an excellent candidate in the first stage to provide initial pixel-level labels of seven geometric classes (namely; sky, support, planar left/right/center, porous and solid). This method extracts color, texture and location features from superpixels, and uses a learning-based boosted decision tree classifier. To enhance the prediction accuracy for sky and support in [16], we develop a 3-class labeling scheme that classifies pixels to three major classes; namely, "support", "vertical" and "sky", where planar left/right/center, porous and solid are merged into one "vertical" mega-class. This 3-class classifier is achieved by integrating segmentation results from [1, 7, 8] with a cascade random forest classifier

[24] and SVM fusion classifier. The performance of our 3-class labeling scheme is 88.7 %, which is better than that in [16] by 2 %. After the 3-class labeling, initial labels of five classes inside the vertical region come directly from the results in [16].

The accuracy of the IPL stage is highly impacted by three factors: (1) a small number of training samples, (2) the weak discriminant power of local features, and (3) lacking of a global scene structure. They are common challenges encountered by all machine learning methods relying on local features with a discriminative model. To overcome these challenges, we turn to a generative model for the whole scene image on top of the discriminative model, which is elaborated in the next subsection.

5.3.3 Global Attributes Extraction (GAE)

In the second GAE stage, we attempt to fill out the 7D binary-valued GAV and find the related information associated with an existing element Fig. 5.9.

Sky and Ground Lines Detection. Sky and ground regions are important ingredients of the geometrical layout of both natural and urban scenes. To infer their existence and correct locations is critical to the task of scene understanding. We develop a robust procedure to achieve this goal as illustrated in Fig. 5.10. Based on initial pixel labels obtained in the first stage, we obtain initial sky and ground lines, which may not be correct due to erroneous initial labels.

To finetune initial sky and ground lines, we extract three cues from the input scene image for sky and ground line validation. They are: (1) the probability edge map of structured edges in [4, 5, 33], (2) the line segments map in [9] and (3) the edge of defocus map in [32]. The sky (or ground) line confidence score is defined as the product of those of all three maps. An example is given in Fig. 5.9, where all three maps have high confidence scores for the sky line but low confidence scores for the ground line. Thus, the fake ground line is removed.

After obtaining the sky and ground lines, we check whether there is any vertical region above the sky line or below the ground line using the 3-class labeling scheme, where the vertical region is either solid or porous. The new 3-class labeling scheme can capture small vertical regions well and, after that, we will zoom into each vertical region to refine its subclass label Fig. 5.11.

Horizon Detection. When the ground plane and the sky plane meet, we see a horizon. This occurs in ocean scenes or scenes with a flat ground. The horizon location helps reason the ground plane, vertical plane and the sky plane. For example, the ground should not be above the horizon while the sky should not be below the horizon. Generally speaking, 3D layout estimation accuracy can be significantly enhanced if the ground truth horizon is available for layout reasoning [16]. Research on horizon estimation from building images was done before, *e.g.*, [2, 14]. That is, it can be inferred by connecting horizontal vanishing points. However, the same technique does not apply to natural scene images where the vanishing point information is lacking.

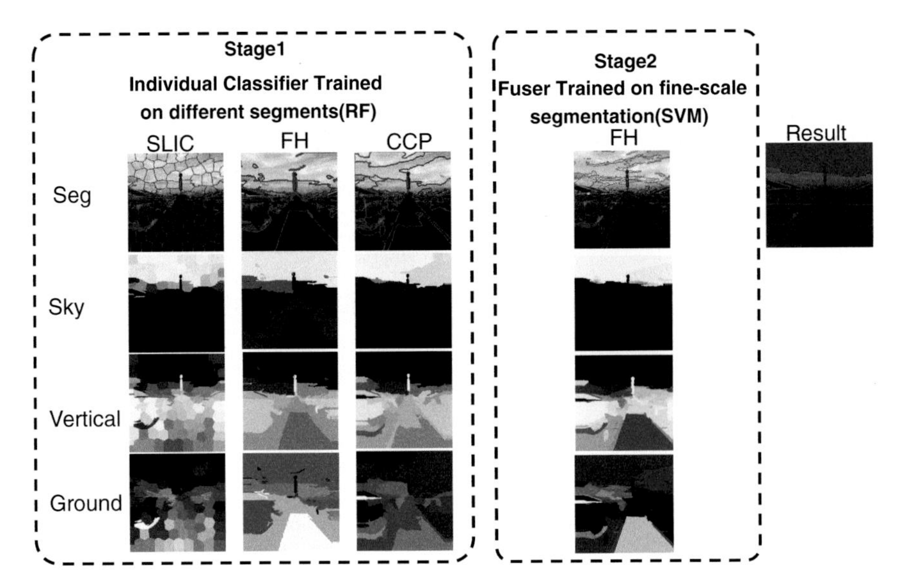

Fig. 5.9 Cascade three class labeling algorithm. *Stage 1* individual random forest classifier trained by segments from SLIC [1], FH [7] and CCP [8]. The gray images show the probability output of individual classifier under different segmentation method and different geometric classes. *Stage 2* cascade the probability output from SLIC [1], FH [7] and CCP [8] in a smaller super-pixel segmentation unit and train a SVM classifier to get the final decision

In our implementation, we use two different methods to estimate the horizon in two different outdoor scenes. For images containing buildings as evidenced by strong vertical line segments, their horizon can be estimated by fitting the horizontal vanishing points [2]. For natural scene images that do not have obvious vanishing points, we propose a horizon estimation algorithm as shown in Fig. 5.12. First, we extract multiple horizontal line segments from the input image using the LSD algorithm [9]. Besides, we obtain the edge probability map based on [4, 5, 33] and use it as well as a location prior to assign a confidence score to each pixel in the line segments. Then, we build a histogram to indicate the probability of the horizon location. Finally, we will select the most likely horizonal line to be the horizon.

After detecting the horizon, we perform layout reasoning to determine the sky and support regions. One illustrative example is given in Fig. 5.12. We can divide the initial labeled segments into two regions. If a segment above (or below) the horizon is labeled as sky (or support), it belongs to the confident region. On the other hand, if a segment above (or below) the horizon is labeled as support (or sky), it belongs to the unconfident region. The green circled region in Fig. 5.12 is labeled as sky due to its white color. Thus, it lies in the unconfident region. There exists a conflict between the local and global decisions. To resolve the conflict, we use the Gaussian Mixture Model (GMM) to represent the color distribution in the confident region above and below the horizon. Then, we conclude that the white color under the horizon can actually be the support (or ground) so that its label can be corrected accordingly.

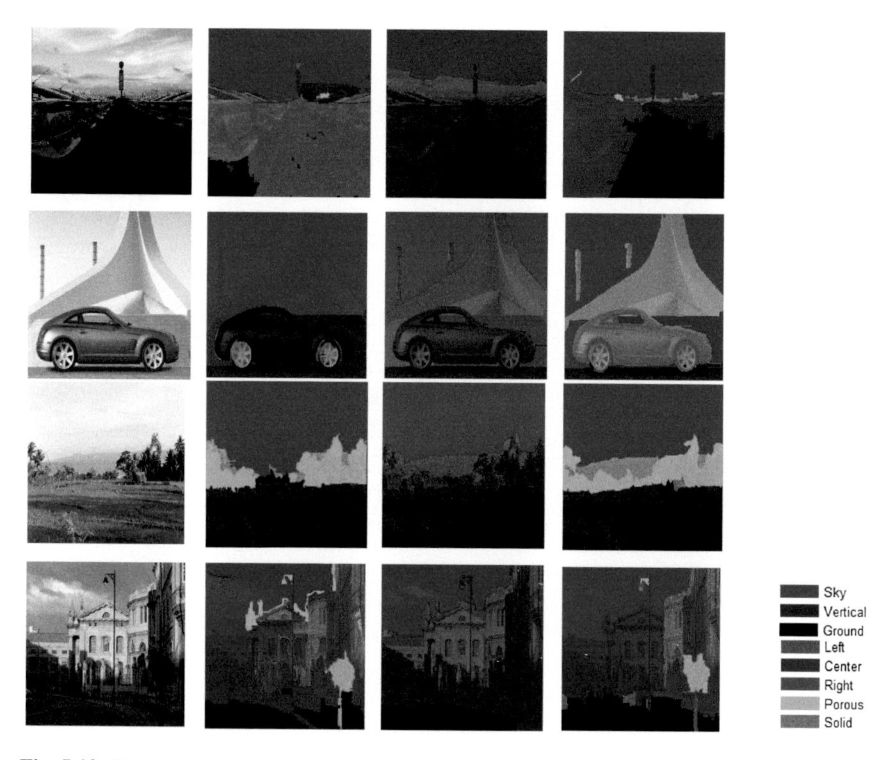

Fig. 5.10 Three class initial label result comparison. From *left* to *right* original image, seven class labeling result from [16], our three class labeling result, ground truth seven class labeling

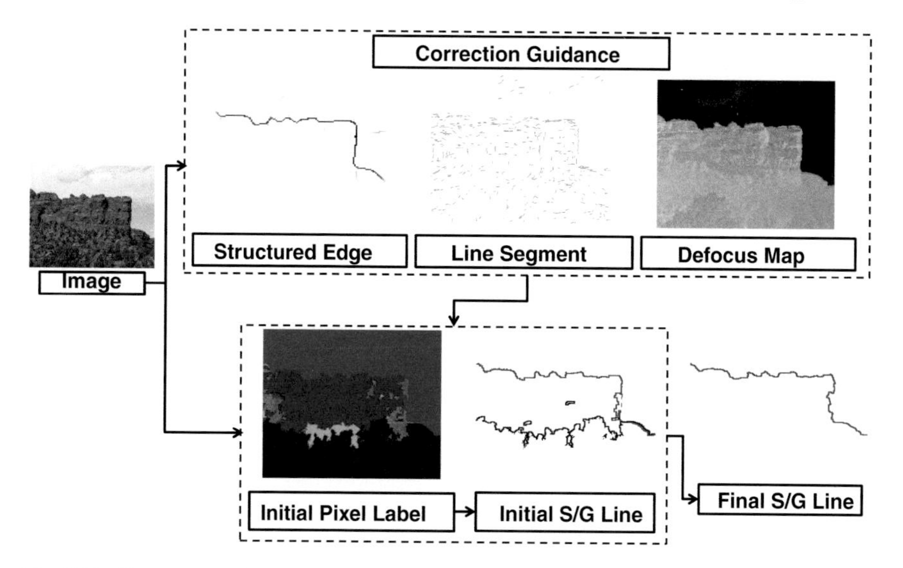

Fig. 5.11 The process of sky/ground line existence validation and location inference

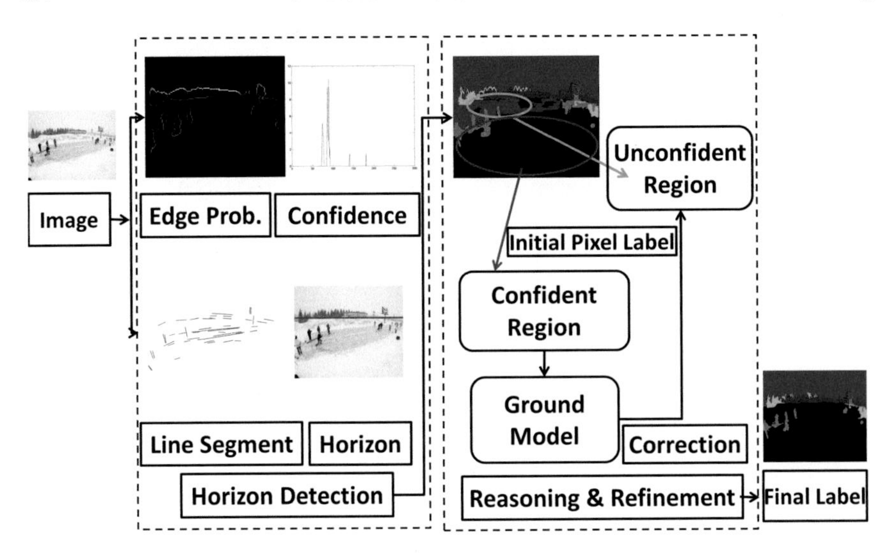

Fig. 5.12 Horizon detection and its application to layout reasoning and label refinement

Planar Surfaces Detection. An important by-product of sky/ground line localization is the determination of planar surface orientation in the vertical region. This is feasible since the shapes of sky and ground lines provide useful cues for planar surface orientation inference in natural or urban scenes. To be more specific, we check the trapezoidal shape fitting scheme (including triangles and rectangles as special cases) for the vertical region, where the top and the bottom of the trapezoidal shape are bounded by the sky and ground lines while its left and right are bounded by two parallel vertical lines or extends to the image boundary.

Three trapezoidal region shape fitting examples are shown in Fig. 5.13. Clearly, different fitting shapes indicate different planar surface orientations. For example, two trapezoidal regions with narrow farther sides indicate an alley scene as shown in the top example. A rectangle shape indicates a front shot of a building as shown in the middle example. The two trapezoidal regions with one common long near sides indicates two building facades with different orientations as given in the bottom example. Thus, the shapes of sky and ground lines offer important cues to planar surface orientations.

Vertical Line Detection. A group of parallel vertical line segments provides a strong indicator of a building structure in scene images. It offers a valuable cue in correcting wrongly labeled regions in building scene images. In our implementation, we use the vertical line percentage in a region as the attribute to generate the probability map for the building region. An example of using vertical lines to correct wrongly labeled regions is illustrated in Fig. 5.14. The top region of the building is wrongly labeled as "sky" because of the strong location and color cues in the initial labeling result. However, the same region has a strong vertical line structure. Since the "sky" region should not have this structure, we can correct its wrong label.

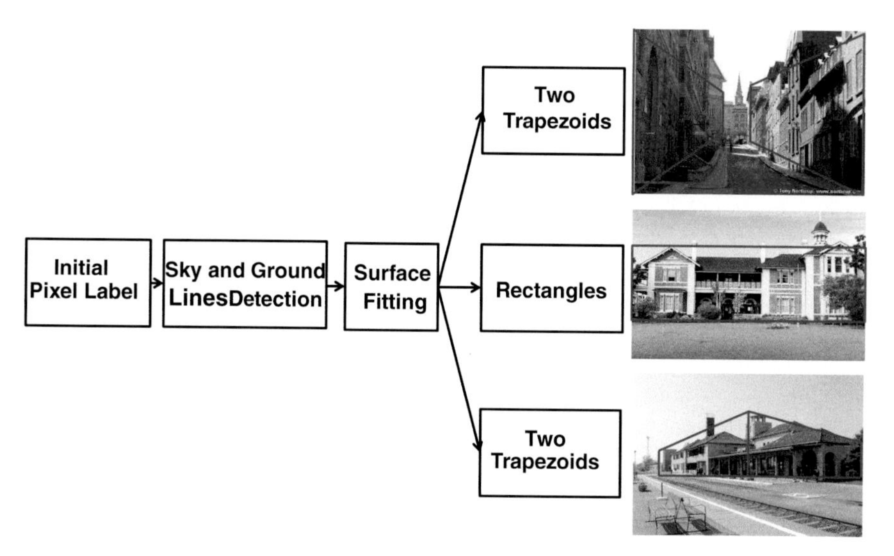

Fig. 5.13 Trapezoidal shape fitting with the *sky line* in the *top* and the *ground line* in the *bottom*

Vanishing Line Detection. A group of vanishing lines offers another global attribute to indicate the surface orientation. In our implementation, we first use the vertical line detection technique to obtain a rough estimate of the left and right boundaries of a building region and then obtain its top and bottom boundaries using the sky line and the ground line. After that, we apply the vanishing line detection algorithm in [15] and use the obtained vanishing line result to adjust the orientation of a building facade. Note that the surface orientation of a planar surface can be obtained from its shape and/or vanishing lines. If it is not a building, we cannot observe vanishing lines and the shape is the only cue. If it is a building we can observe both its shape and vanishing lines. The two cues are consistent with each other based on our experience. Two examples of using the shape and vanishing lines of a building region to correct erroneous initial labels are given in Fig. 5.15. The initial surface orientation provided by the IPL stage contains a large amount of errors due to the lack of global view. They are corrected using the global attributes.

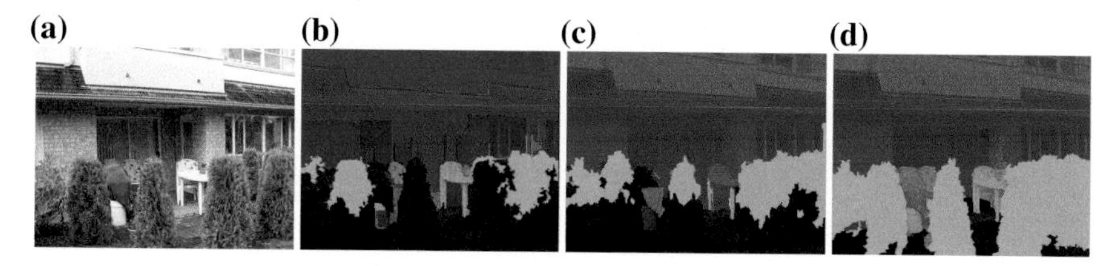

Fig. 5.14 An example of *vertical line* detection and correction. **a** Original image **b** Initial label **c** Refined label **d** Ground truth

Solid Detection. The non-planar solid class is typically composed by foreground objects (such as people, car, etc.) rather than the background scene. The bottom of an object either touches the ground or extends to the image bottom boundary. For example, objects (say, pedestrians and cars) may stand on the ground in front of the building facade in an urban scene while objects are surrounded by the ground plane in a typical natural scene. Object detection is an active research field by itself. Actually, the object size has an influence on the scene content, i.e. whether it is an object-centric or scene-centric image. The object size is big in an object-centric image. It occupies a large portion of the image, and most of the background is occluded. The scene layout problem is of less interest since the focus is on the object rather than the background. In contrast, there is no dominant object in a scene-centric image and the scene layout problem is more significant. On one hand, it is better to treat the object detection problem independently from scene analysis. On the other hand, a scene image may still contain some objects while the background scene is occluded by them. For examples, the existence of objects may occlude some part of sky and ground lines, thereby increasing the complexity of scene layout layout estimation. So, in order to simplify the scene understanding problem, it is desired to identify and remove the objects at first.

In our implementation, we apply two object detectors in [6, 10]—the person detector and the car detector. We first obtain the bounding boxes of detected objects, then apply the grab cut [30] method to get their exact contours. Two examples are shown in Fig. 5.16. To achieve a better layout estimation, detected and segmented objects are removed from the scene image to allow more reliable sky and ground line detection.

Porous Detection. As shown in the bottom subfigure of Fig. 5.17, the mountain region, which belongs to the solid class, is labeled as porous by mistake. This is due

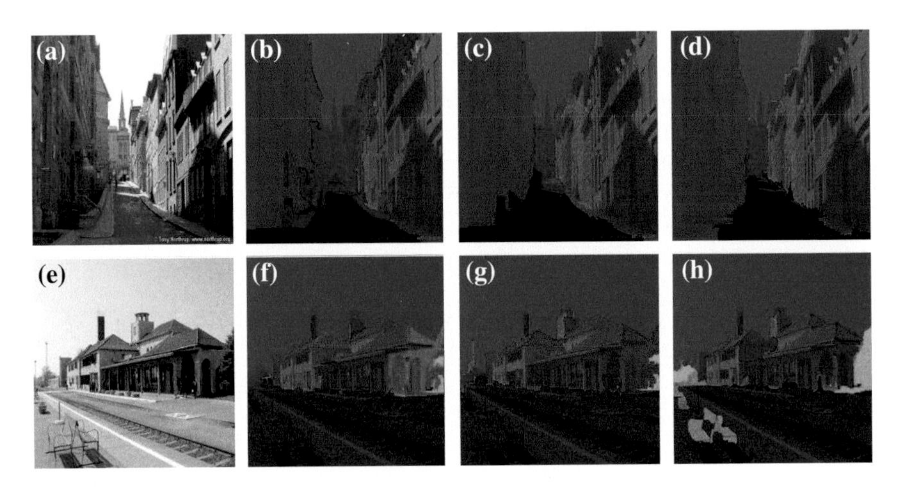

Fig. 5.15 Examples of surface orientation refinement using the shape and the *vanishing line* cues. **a** Original image **b** Initial label **c** Refined label **d** Ground truth

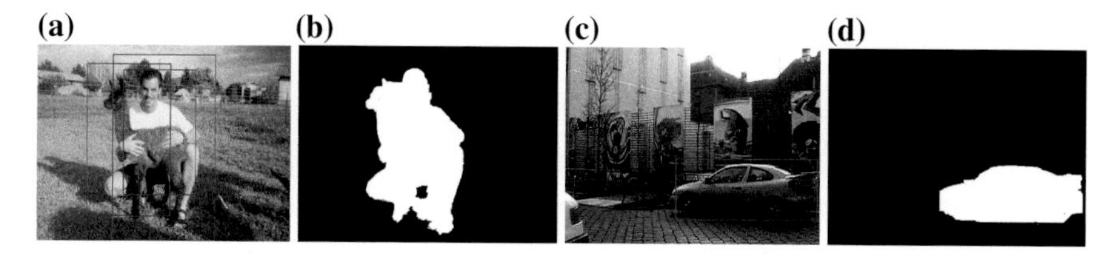

Fig. 5.16 Examples of obtaining the object mask using the person and car object detectors and the grab cut segmentation method. **a** Person detector bounding boxes result **b** Person segmentation result **c** Car detector bounding box result **d** Car segmentation result

to the fact that the superpixel segmentation method often merges the porous region with its background. Thus, it is difficult to split them in the classification stage. To overcome this difficulty, we add the contour randomness feature from the structured edge result [4, 5, 33] to separate the porous and the solid regions. By comparing the two examples in Fig. 5.17, we see that there exist irregular contours inside the true porous region (trees) but regular contours inside the solid region (mountain). In our implementation, we double check regions labeled by solid or porous initially and use the contour smoothness/randomness to separate solid/porous regions.

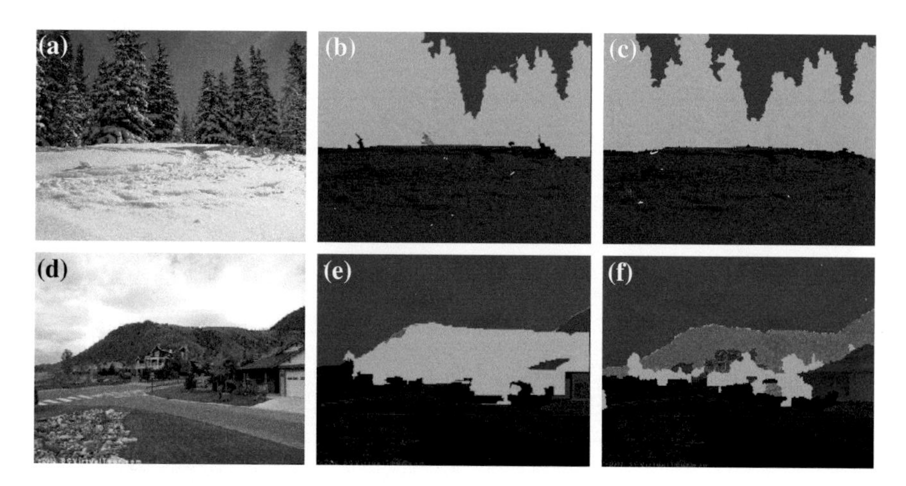

Fig. 5.17 Two examples of initially labeled porous regions where the *top* (trees) is correctly labeled while the *bottom* (mountain) is wrongly labeled. **a** Original image **b** Initial pixel label **c** Ground truth **d** Original image **e** Initial pixel label **f** Ground truth

5.3.4 Layout Reasoning and Label Refinement (LR2)

The binary-valued 7D GAV can characterize a wide range of scene images. Some ideas on layout reasoning were already discussed in the previous subsection, and they will not be repeated. Here, we attempt to integrate all attributes together and propose a geometric layout framework from the simplest setting to the general setting as shown in Fig. 5.18.

The simplest case of our interest is a scene with a horizon as shown in Fig. 5.18a. This leads to sky and support two regions and the support could be ocean or ground. The existence of occluders will make this case slightly more complicated. The sky may have two occluders—solid (e.g., balloon, bird, airplane, kite, parashoot) and porous (e.g., tree leaves near the camera). The ocean/ground may also have occluders—solid (e.g., boat in the ocean, fence/wall near the camera) and porous (e.g. bushes near the camera). Although there are rich scene varieties, we can remove segments labeled with planar center/left/right in the simplest case confidently.

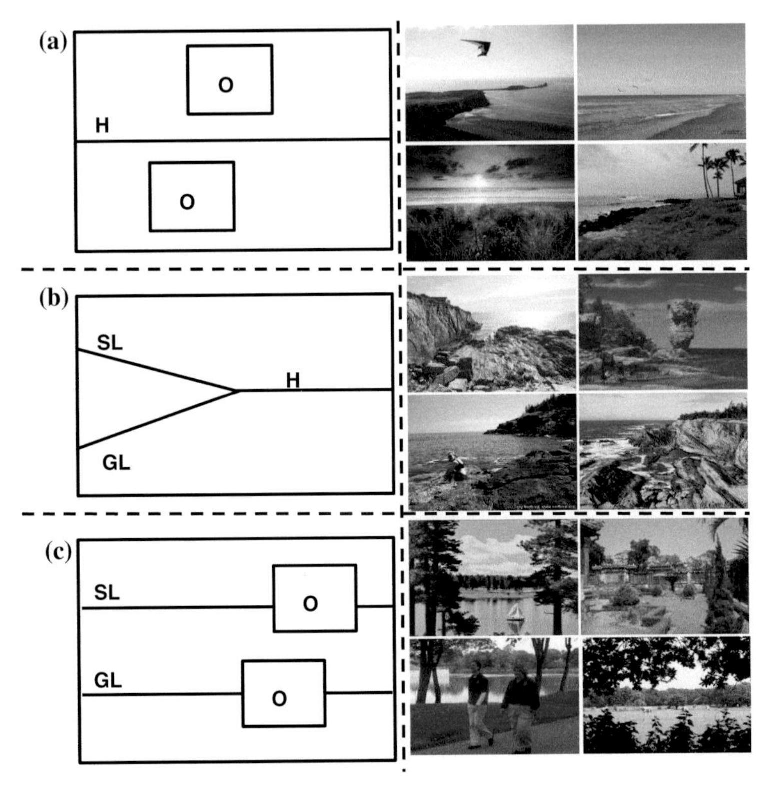

Fig. 5.18 A proposed framework for geometric layout reasoning: **a** the simplest case, **b** the background scene only, and **c** a general scene, where SL, GL, H, O denote the *sky line*, the *ground line*, the *horizon* and the occluder, respectively

The horizon will be split into individual sky and ground lines if there exist large planar surfaces that either completely or partially block the horizon as shown in Fig. 5.18b. It is a background scene only, if there is no occluder. The large planar surfaces can be buildings in urban scenes and mountains in natural scenes. Generally speaking, any region between the sky and ground lines is a strong candidate for planar surfaces. We can further classify them into planar left/right/center based on vanishing lines or the outer contour of the surface region.

However, a general scene may have all kinds of solid/porous occluders (e.g. people, car, trees, etc.) in front of a background scene as shown in Fig. 5.18c. The occluders may block the ground line, the planar surface and other solid/porous objects to make layout inference more difficult. If we can have powerful object detectors to remove these occluders, the scene layout problem can be greatly simplified.

Finally, if the initial label is consistent with the layout reasoning result, no label refinement is needed. Otherwise, we will refine labels based on the global layout reasoning.

5.4 Experimental Results

In contrast to the object detection problem, very few datasets are available for geometric layout algorithms evaluation. The dataset in [16] is the largest benchmarking dataset in the geometric layout research community. It consists of 300 images of outdoor scenes with labeled ground truth. The first 50 images are used for training the surface segmentation algorithm as done in previous work [11, 16]. The remaining 250 images are used for evaluation. We follow the same procedure in our experiments. Also, we use the initial labeling result from [16] in the IPL stage. To test our generalized global attributes and reasoning system, we use the same 250 testing images in our experiment.

It is not a straightforward task to label the ground truth for outdoor scene images in this dataset. It demands subjects to reason the functionalities of different surfaces in the 3D real world (rather than recognizing objects in images only), e.g., where is the occlusion boundary? what is the surface orientation? whether there is depth difference? etc. To get a consistent ground truth, subjects should be very well trained. Note that we do not agree with the labeled ground truth of several images, and will report them in the supplemental material. In this section, we will present both qualitative and quantitative evaluation results and analyze several poorly labeled images.

Qualitative Analysis. Figure 5.19 shows eight representative images (the first column) with their labeled ground truths (the last column). More visual results are shown in Figs. 5.20 and 5.21. We show results obtained by Hoiem et al. [16] (the second column), Gupta et al. [11] (the third column), the proposed GAL (the fourth column). We call the methods proposed in Hoiem et al. [16] and Gupta et al. [11] the H-method and the G-method, respectively, in the following discussion for convenience.

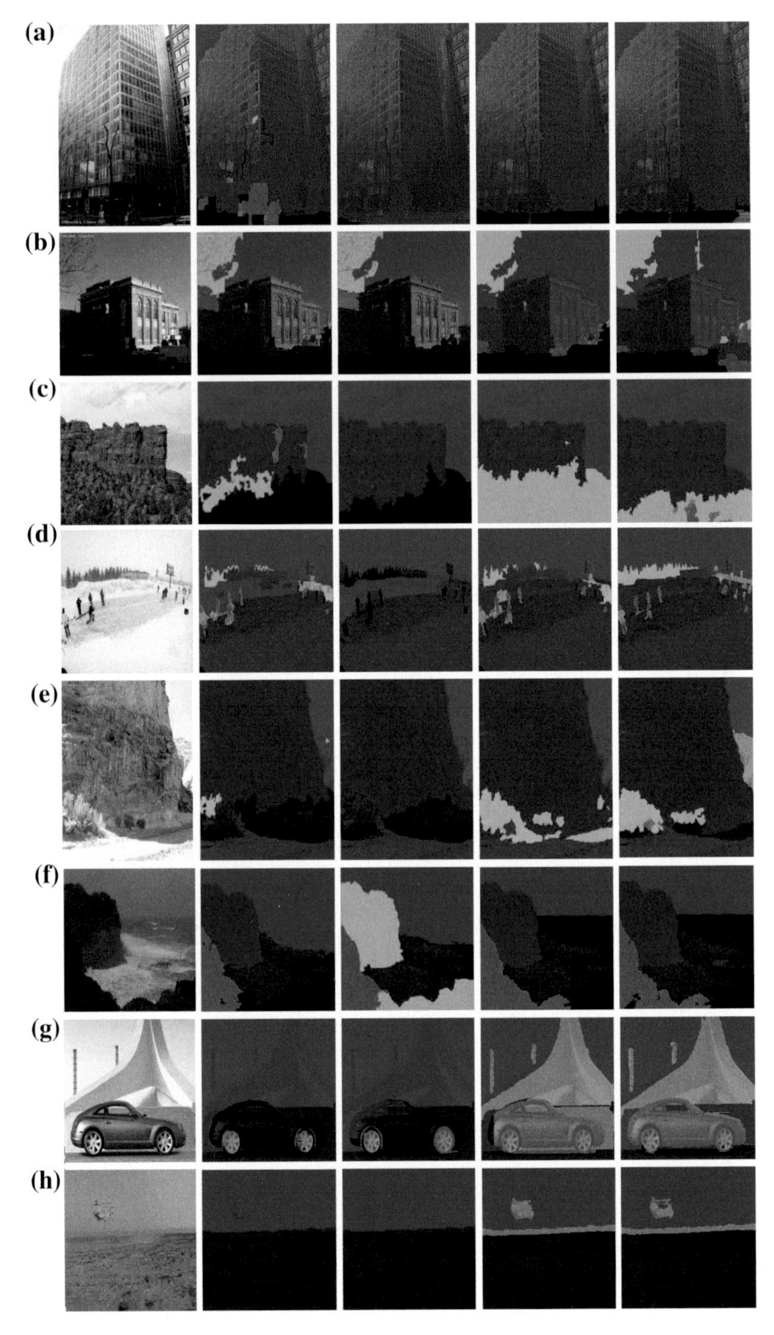

Fig. 5.19 Qualitative comparisons of three geometric layout algorithms (from *left* to *right*): the original Image, Hoiem et al. [16], Gupta et al. [11], GAL and the ground truth. The surface layout color codes are: *magenta* (planar *left*), *dark blue* (planar *center*), *red* (planar *right*), *green* (non-planar porous), *gray* (non-planar solid), *light blue* (sky), *black* (support)

Fig. 5.20 Qualitative comparisons of three geometric layout algorithms (from *left* to *right*): the original Image, Hoiem et al. [16], Gupta et al. [11], GAL and the ground truth. The surface layout color codes are: *magenta* (planar *left*), *dark blue* (planar *center*), *red* (planar *right*), *green* (non-planar porous), *gray* (non-planar solid), *light blue* (sky), *black* (support)

Fig. 5.21 Qualitative comparisons of three geometric layout algorithms (from *left* to *right*): the original Image, Hoiem et al. [16], Gupta et al. [11], GAL and the ground truth. The surface layout color codes are: *magenta* (planar *left*), *dark blue* (planar *center*), *red* (planar *right*), *green* (non-planar porous), *gray* (non-planar solid), *light blue* (sky), *black* (support)

In Fig. 5.19a, the building is tilted and its facade contains multiple surfaces with different orientations. Because of the similarity of the local visual pattern (color and texture), the H-method is confused by upper and lower parts in the planar-left facade, and assigns them to planar left and right two opposite oriented surfaces. This error can be corrected by the global attribute (i.e. building's vertical lines). The G-method loses the detail of surface orientation in the upper left corner so that the sky and the building surface are combined into one single planar left surface. It also loses the ground support region. These two errors can be corrected by sky-line and ground-line detection. GAL infers rich semantic information about the image from the 7D GAV. Both the sky and the ground exist in the image. Furthermore, it is a building facade due to the existence of parallel vertical lines. It has two strong left and right vanishing points so that it has two oriented surfaces—one planar left and the other planar right. There is a problem remaining in the intersection region of multiple buildings in the right part of the image. Even humans may not have an agreement on the number of buildings in that area (two, three, or four?). The ground truth also looks strange. It demands more unambiguous global information for scene layout inference.

For Fig. 5.19b, all three layout methods offer good results. The ground truth has two problems. One is in the left part of the image. It appears to be more reasonable to label the building facade to "planar left" rather than "planar center". Another is in the truck car region. The whole truck should have the same label; namely, gray (non-planar solid). This example demonstrates the challenging nature of the geometric layout problem. Even human can make mistakes easily.

Figure 5.19c–f show the benefit of getting an accurate sky line and/or an accurate ground line. One powerful technique to find the sky line and the ground line is to use defocus estimation. Once these two lines are detected, the inference becomes easier so that GAL yields better results.

Being constrained by the limited model hypothesis, the G-method chooses to have the ground (rather than the porous) to support the physical world. However, this model does not apply to the image in (c), images with bushes in the bottom as an occluder, etc. Also, the ground truth in (c) is not consistent in the lower region. The result obtained by GAL appears to be more reasonable.

Figure 5.19d was discussed earlier. Both the H-method and the G-method labeled part of the ground as the sky by mistake due to color similarity. GAL can fix this problem using horizon line detection. All three methods work well for Fig. 5.19e. GAL performs slightly better due to the use of the horizon detection and label refinement stage. The right part of the image is too blurred to be of interest. GAL outperforms the H-method and the G-method due to an accurate horizon detection for Fig. 5.19f. Figure 5.19g demonstrates the power of the sky line detection and object (car) detector. Figure 5.19h demonstrates the importance of an accurate sky line detection. We also compare the "pop-up" view of the H-method and GAL in Fig. 5.22 using a 3D view rendering technique given in [19]. It is clear that GAL offers a more meaningful result. More results are shown in Fig. 5.23.

We have the following conclusion from the above discussion. First, the local-patch-based machine learning algorithm (e.g., the H-method) lacks the global information for 3D layout reasoning. Second, it is difficult to develop complete outdoor

Fig. 5.22 Comparison of 3D rendered views based on geometric labels from the H-method (*left*) and GAL (*right*). **a** "Pop-up" by H-method **b** "Pop-up" by Proposed GAL

scene models to cover a wide diversity of outdoor images (e.g., the G-method). GAL attempts to find key global attributes from each outdoor image. As long as these global attributes can be determined accurately, we can combine the local properties (in form of initial labels) and global attributes for layout reasoning and label refinement. The proposed methodology can handle a wider range of image content. As illustrated above, it can provide more satisfactory results to quite a few difficult cases that confuses prior art such as the H-method and the G-method.

Quantitative Analysis. For quantitative performance evaluation, we use the metric used in [11, 19] by computing the percentage of overlapping of labeled pixel and ground truth pixel (called the pixel-wise labeling accuracy). We use the original ground truth given in [17] for fair comparison between all experimental results reported below (with only one exception which will be clearly stated).

We first compare the labeling accuracy of the H-method, the G-method and GAL for seven geometric classes individually in Fig. 5.24. We see clearly that GAL outperforms the H-method in all seven classes by a significant margin. The gain ranges from 1.27 % (support) to more than 25 % (planar left and right). GAL also outperforms the G-method in 6 classes. The gain ranges from 0.81 % (sky) to 14.9 % (solid). The G-method and GAL have comparable performance with respect to the "planar center" class. All methods do well in sky and ground labeling. The solid class is the most challenging one among all seven classes since it has few common geometric structures to exploit.

Fig. 5.23 Comparison of 3D rendered views based on geometric labels from the H-method (*left*) and GAL (*right*). **a** "Pop-up" by H-method **b** "Pop-up" by Proposed GAL

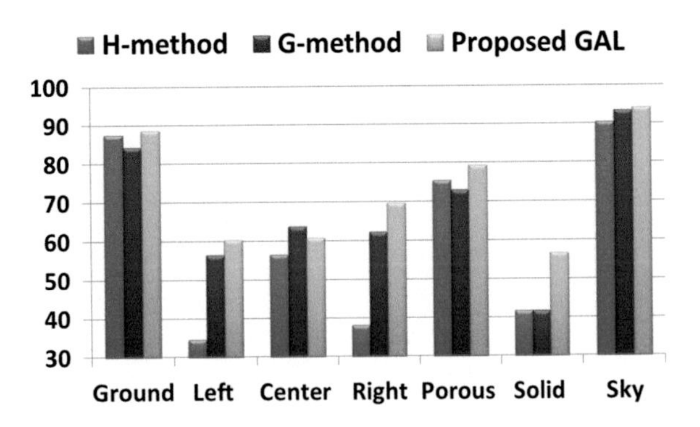

Fig. 5.24 Comparison of labeling accuracy for the H-method, the G-method and GAL with respect to seven individual labels

The complete performance benchmarking (in terms of pixel-wise labeling accuracy) of five different methods are shown in Table 5.1. The L-method in [26] and the P-method in [28] were dedicated to building facade labeling and their performance were only evaluated on subsets of the full dataset in [17]. The L-method in [26] use the subset which contains 100 images images where the ground truth of both occlusion boundaries and surface orientation are provided [16]. P-method in [28] use the subset which contains 55 building images.

We use B1, B2, F and F/R to denote the subset used in [26], the subset used in [28], the full dataset and the full dataset with relabeled ground truth in the table, respectively. The two numbers within the parenthesis (7 and 5) denote results for all "seven" classes and for the "five" vertical classes (i.e., excluding sky and support), respectively. For the F(7) column in Table 5.1, we compare the performance for all seven classes in the full set. Accurate labeling for this dataset is actually very challenging as pointed out in [22, 29]. This is also evidenced by the slow performance improvement over the last seven years—a gain of 2.35 % from the H-method to the G-method. GAL offers another gain of 4.74 % over the G-method [11], which is significant. For the B1(5) column, we compare the performance for five classes in the building subset which contains 100 images. GAL outperforms the L-method by 3.04 %. For the B2(7) column, we compare the performance against the building

Table 5.1 Comparison of the averaged labeling accuracy (%) of five methods with respect to the building subset (B), the full set (F) and the full set with relabeled ground truth (F/R), where 7 and 5 mean all seven classes and the five classes belonging to the vertical category, respectively

	Dataset (Class No.)				
	B1(5)	B2(7)	F(5)	F(7)	F/R(7)
H-method [16]	N/A	72.87	68.80	72.41	N/A
G-method [11]	N/A	73.59	73.72	74.76	N/A
L-method [26]	76.34	N/A	N/A	N/A	N/A
P-method [28]	N/A	74.82	N/A	N/A	N/A
Proposed GAL	79.38	81.24	76.17	79.50	80.05

subset which contains 55 images. GAL outperforms the H-method, the G-method and the P-method by 8.37, 7.65 and 6.42 %, respectively. For the F(5) column, we compare the performance for five classes in the full set. GAL outperforms the H-method, the G-method by 7.37 and 2.45 %. Finally, for the F/R(7) column, We show the labeling accuracy of GAL against the modified ground truth, shown in Fig. 5.25. The labeling accuracy can go up to 80.05 %.

Error Analysis. Three exemplary images that have large labeling errors are shown in Fig. 5.26. Figure 5.26a is difficult due to the complexity in the middle region of multiple houses. We feel that the labeled "planar center" result by GAL for this region is still reasonable although it is not as refined as that offered by the ground truth. Figure 5.26b is one of the most challenging scenes in the dataset since even humans may have disagreement. The current ground truth appears to be too complicated to be useful. GAL can find the ground line but not the sky line due to the tree texture in the top part of the image. GAL made a labeling mistake in porous in Fig. 5.26c due to dominant texture in the corresponding region. We need a better porous detector

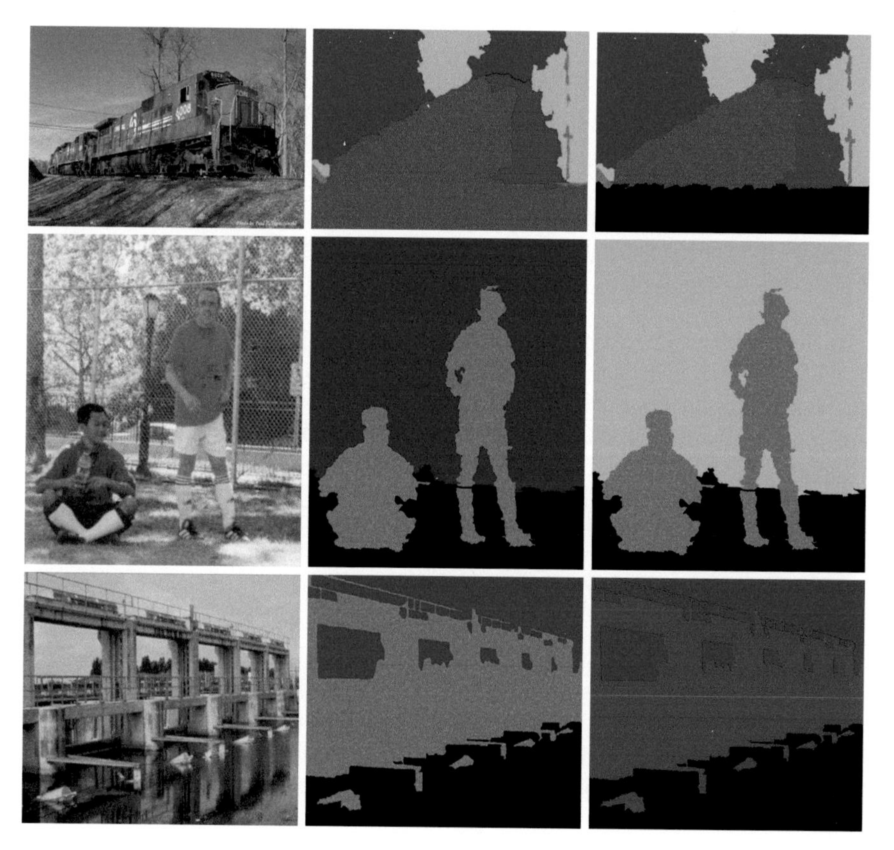

Fig. 5.25 Three modified ground truth. From *left* to *right* **a** Original image **b** Ground truth in [16] **c** Relabeled ground truth

Fig. 5.26 Error analysis of the proposed GAL system with three exemplary images (one example per row and from *left* to *right*) the original image, the labeled result of GAL and the ground truth

to fix this problem. The global attributes will not be able to help much. There is a planar right region in the ground truth. Although this label is accurate to human eyes by looking at the local surface, the transition from support (or ground) to planar right is not natural. It could be an alternative to treat the whole region in the bottom part as support. As shown in this image, it is extremely difficult to find an exhaustive set of scene models to fit all different situations.

5.5 Summary

A novel GAL geometric layout labeling system for outdoor scene images was proposed in this work. GAL exploits both local and global attributes to achieve higher accuracy and it offers a major advancement in solving this challenging problem. Its performance was analyzed both qualitatively and quantitatively. Besides, several error cases were studied to reveal the limitations of the proposed GAL system. Clearly, there are still many interesting problems remaining, including the development of better global attributes extraction tools, its integration with object detectors, and the design of more powerful and more general inference rules. Furthermore, due to major differences between indoor and outdoor scene images, the key global attributes of indoor scene images will be different from those of outdoor scene images. So, more research needs to be done to develop a GAL system for indoor scenes. Future research is planned to determine important global attributes of indoor scene images and to study their geometric layout.

References

1. Achanta, R., Shaji, A., Smith, K., Lucchi, A., Fua, P., Susstrunk, S.: Slic superpixels compared to state-of-the-art superpixel methods. IEEE Trans. Pattern Anal. Mach. Intell. **34**(11), 2274–2282 (2012)
2. Barinova, O., Lempitsky, V., Tretiak, E., Kohli, P.: Geometric image parsing in man-made environments. Computer VisionECCV 2010, pp. 57–70. Springer (2010)
3. Choi, W., Chao, Y.W., Pantofaru, C., Savarese, S.: Understanding indoor scenes using 3d geometric phrases. In: EEE Conference on Computer Vision and Pattern Recognition (CVPR), 2013 I, pp. 33–40. IEEE (2013)
4. Dollár, P., Zitnick, C.L.: Structured forests for fast edge detection. In: IEEE International Conference on Computer Vision (ICCV), 2013, pp. 1841–1848. IEEE (2013)
5. Dollár, P., Zitnick, C.L.: Fast edge detection using structured forests (2014)
6. Felzenszwalb, P.F., Girshick, R.B., McAllester, D., Ramanan, D.: Object detection with discriminatively trained part based models. IEEE Trans. Pattern Anal. Mach. Intell. **32**(9), 1627–1645 (2010)
7. Felzenszwalb, P.F., Huttenlocher, D.P.: Efficient graph-based image segmentation. Int. J. Comput. Vis. **59**(2), 167–181 (2004)
8. Fu, X., Wang, C.Y., Chen, C., Wang, C., Kuo, C.C.J.: Robust image segmentation using contour-guided color palettes (2015)
9. von Gioi, R.G., Jakubowicz, J., Morel, J.M., Randall, G.: Lsd: a fast line segment detector with a false detection control. IEEE Trans. Pattern Anal. Mach. Intell. **4**, 722–732 (2008)
10. Girshick, R.B., Felzenszwalb, P.F., McAllester, D.: Discriminatively trained deformable part models, release 5. http://people.cs.uchicago.edu/rbg/latent-release5/
11. Gupta, A., Efros, A.A., Hebert, M.: Blocks world revisited: image understanding using qualitative geometry and mechanics. Computer VisionECCV 2010, pp. 482–496. Springer (2010)
12. Gupta, A., Hebert, M., Kanade, T., Blei, D.M.: Estimating spatial layout of rooms using volumetric reasoning about objects and surfaces. In: Advances in Neural Information Processing Systems, pp. 1288–1296 (2010)
13. Gupta, A., Satkin, S., Efros, A., Hebert, M.: From 3d scene geometry to human workspace. In: 2011 IEEE Conference on Computer Vision and Pattern Recognition (CVPR), pp. 1961–1968. IEEE (2011)
14. Han, F., Zhu, S.C.: Bottom-up/top-down image parsing with attribute grammar. IEEE Trans. Pattern Anal. Mach. Intell. **31**(1), 59–73 (2009)
15. Hedau, V., Hoiem, D., Forsyth, D.: Recovering the spatial layout of cluttered rooms. In: 2009 IEEE 12th International Conference on Computer vision, pp. 1849–1856. IEEE (2009)
16. Hoiem, D., Efros, A., Hebert, M.: Closing the loop in scene interpretation. In: IEEE Conference on Computer Vision and Pattern Recognition, 2008. CVPR 2008, pp. 1–8. IEEE (2008)
17. Hoiem, D., Efros, A., Hebert, M., et al.: Geometric context from a single image. In: Tenth IEEE International Conference on Computer Vision, 2005. ICCV 2005, vol. 1, pp. 654–661. IEEE (2005)
18. Hoiem, D., Efros, A.A., Hebert, M.: Automatic photo pop-up. ACM Trans. Graph. (TOG) **24**(3), 577–584 (2005)

19. Hoiem, D., Efros, A.A., Hebert, M.: Recovering surface layout from an image. Int. J. Comput. Vis. **75**(1), 151–172 (2007)
20. Hoiem, D., Efros, A.A., Hebert, M.: Recovering occlusion boundaries from an image. Int. J. Comput. Vis. **91**(3), 328–346 (2011)
21. Koutsourakis, P., Simon, L., Teboul, O., Tziritas, G., Paragios, N.: Single view reconstruction using shape grammars for urban environments. In: 2009 IEEE 12th International Conference on Computer Vision, pp. 1795–1802. IEEE (2009)
22. Lazebnik, S., Raginsky, M.: An empirical bayes approach to contextual region classification. In: IEEE Conference on Computer Vision and Pattern Recognition, 2009. CVPR 2009, pp. 2380–2387. IEEE (2009)
23. Lee, D.C., Hebert, M., Kanade, T.: Geometric reasoning for single image structure recovery. In: IEEE Conference on Computer Vision and Pattern Recognition, 2009. CVPR 2009, pp. 2136–2143. IEEE (2009)
24. Liaw, A., Wiener, M.: Classification and regression by randomforest. R news **2**(3), 18–22 (2002)
25. Liu, B., Gould, S., Koller, D.: Single image depth estimation from predicted semantic labels. In: 2010 IEEE Conference on Computer Vision and Pattern Recognition (CVPR), pp. 1253–1260. IEEE (2010)
26. Liu, X., Zhao, Y., Zhu, S.C.: Single-view 3d scene parsing by attributed grammar. In: 2014 IEEE Conference on Computer Vision and Pattern Recognition (CVPR), pp. 684–691. IEEE (2014)
27. Mobahi, H., Zhou, Z., Yang, A.Y., Ma, Y.: Holistic 3d reconstruction of urban structures from low-rank textures. In: 2011 IEEE International Conference on Computer Vision Workshops (ICCV Workshops), pp. 593–600. IEEE (2011)
28. Pan, J., Hebert, M., Kanade, T.: Inferring 3d layout of building facades from a single image. In: Proceedings of the IEEE Conference on Computer Vision and Pattern Recognition, pp. 2918–2926 (2015)
29. Ramalingam, S., Kohli, P., Alahari, K., Torr, P.H.: Exact inference in multi-label crfs with higher order cliques. In: IEEE Conference on Computer Vision and Pattern Recognition, 2008. CVPR 2008, pp. 1–8. IEEE (2008)
30. Rother, C., Kolmogorov, V., Blake, A.: Grabcut: Interactive foreground extraction using iterated graph cuts. ACM Trans. Graph. (TOG) **23**(3), 309–314 (2004)
31. Saxena, A., Sun, M., Ng, A.Y.: Make3d: learning 3d scene structure from a single still image. IEEE Trans. Pattern Anal. Mach. Intell. **31**(5), 824–840 (2009)
32. Zhuo, S., Sim, T.: Defocus map estimation from a single image. Pattern Recogn. **44**(9), 1852–1858 (2011)
33. Zitnick, C.L., Dollár, P.: Edge boxes: locating object proposals from edges. In: Computer Vision-ECCV 2014, pp. 391–405. Springer (2014)

Chapter 6
Conclusion and Future Work

Keywords Big visual data analysis · Indoor/outdoor classification · Outdoor scene classification · Geometric labeling · Classification/labeling accuracy · Robustness · EDF · CSS · GAL

Visual data analysis has attracted a wide range of interests and applications in the computer vision field. With "big data" becoming popular in recent years, many researchers move their focus on big visual data analysis. As one of the most challenging topics, scene understanding for big visual data is having more and more attentions. In this book, we had discussed several traditional visual data analysis approaches for three scene classification problems, indoor/outdoor classification and outdoor scene categorization and outdoor scene geometric labeling. When met with big visual data, these traditional approaches lose both accuracies and robustness on classification performance due to data diversities and abundances. In this book, in order to understand the traditional approaches and tangle their problems when applied to big visual data, three novel algorithms are proposed.

Expert Decision Fusion (EDF) is a advanced structured system targeting at classification improvement. It integrates multiple classification experts' capabilities by data grouping and decision stacking with a careful designed structured system. The experiments, conducted on a big visual dataset, SUN, show the efficiencies of this system on large-scale image classification for indoor/outdoor scene classification.

Coarse Semantic Segmentation (CSS) focuses on achieving a reasonable scale of local learning visual units, which is a basic problem faced by all visual analysis solutions. With the help of the most powerful image segmentation algorithm and a well-designed segmental labeling system, CSS is able to robustly prepare clean and significant learning units to accurately represent scene characteristics. The state-of-the-art performance achieved by CSS in outdoor scene classification clearly demonstrates its capabilities and efficiencies.

Global-attributes Assisted Labeling (GAL) takes the advantages of global visual reasoning procedures in the human visual system. GAL explores the global rules that human brain used to construct rough structures from a single view. This rules are generalized in GAL mathematically with significant visual cluesautomatically

C. Chen et al., *Big Visual Data Analysis*,
SpringerBriefs in Signal Processing, DOI 10.1007/978-981-10-0631-9_6

detected. GAL enables the reasoning on a broad range of scene image types, which are not restricted to city landscapes, where previous works perform accurately. As a general framework, GAL can also be easily extended to indoor scene geometrical labeling or even more challenging scene understanding problems.

For the proposed EDF, CSS and GAL approaches to be applicable in real-world applications efficiently and robustly, more future researches are needed along the following two major directions.

Accuracies:
In Chap. 3, we see that EDF has already achieve 91.15 % on correct classification rate for indoor/outdoor scene classification. However, in real-world applications, 91.15 % are far from sufficient. As we analyzed in Chap. 3, EDF is built on proposed individual experts. These experts' capabilities define the bottleneck for EDF's performance. To have more accurate classification performance, more effort should be spent on the experts designs. Besides, we also observe a increase in classification accuracy when more data is used in our experiments. Since SUN only owns 108,754 images, we have no idea of the converged performance with more available data. As a future research direction, we will also find more images to further test EDF's capabilities with even larger datasets.

Robustness:
In Chaps. 4 and 5, we use CSS and GAL to achieve results on both outdoor scene classification and geometric labeling, where they are far superior than previous works. However, limited by the availabilities of the labeled data, we have to admit that our success need more approval by applying to wild and big visual datasets. The robustness of these two approaches will be definitely a big challenges for us to consider, since they are well designed under the observations of all the available data. In Chap. 4, we only considered 8 outdoor categories. Although they are already representative and covers the general types of scenes in the real world, we cannot prove the true robustness of CSS when applied to complicate outdoor scenes. We have similar worries with the performance of GAL as well. As a result, more work should be completed to further fine-tone the design of the CSS and the GAL system by adding new generalized labels and constraints correspondingly, in order to make them applicable to real-world data with acceptable robustness.